T0139318

MOLECULAR BIOLOGY
INTELLIGENCE
UNIT

Omics Technologies in Cancer Biomarker Discovery

Xuewu Zhang, PhD
South China University of Technology
College of Light Industry and Food Sciences
Guangzhou, China

LANDES BIOSCIENCE
AUSTIN, TEXAS
USA

Omics Technologies in Cancer Biomarker Discovery

Molecular Biology Intelligence Unit

Landes Bioscience

Printed in the USA.

Please address all inquiries to the publisher:
Landes Bioscience, 1806 Rio Grande, Austin, Texas 78701, USA
Phone: 512/ 637 6050; Fax: 512/ 637 6079
www.landesbioscience.com

The chapters in this book are available in the Madame Curie Bioscience Database.
http://www.landesbioscience.com/curie

ISBN: 978-1-58706-339-8

While the authors, editors and publisher believe that drug selection and dosage and the specifications and usage of equipment and devices, as set forth in this book, are in accord with current recommendations and practice at the time of publication, they make no warranty, expressed or implied, with respect to material described in this book. In view of the ongoing research, equipment development, changes in governmental regulations and the rapid accumulation of information relating to the biomedical sciences, the reader is urged to carefully review and evaluate the information provided herein.

Library of Congress Cataloging-in-Publication Data

Zhang, Xuewu, 1963-
 Omics technologies in cancer biomarker discovery / Xuewu Zhang.
   p. ; cm. -- (Molecular biology intelligence unit)
 Includes bibliographical references and index.
 ISBN 978-1-58706-339-8
 1. Tumor markers. 2. Genomics. 3. Tumor proteins. I. Title. II. Series: Molecular biology intelligence unit.
 [DNLM: 1. Tumor Markers, Biological. 2. Genomics--methods. 3. Neoplasm Proteins--diagnostic use. QZ 241]
 RC270.3.T84Z43 2011
 616.99'4075--dc22
                      2011001462

About the Editor...

XUEWU ZHANG received his BSc in Mathematics in 1983 at Hunan Normal University, China. He got his MSc degree in Ecology in 1990 and PhD degree in Entomology in 1993, both from Zhongshan University, China. Then he worked at the Food Engineering Research Center of State Education Ministry, Zhongshan University, China. After May of 1998, he was invited to do postdoctoral research at several universities, including The University of Hong Kong, University of British Columbia (Canada), University of Manitoba (Canada) and University of California at Los Angeles (USA). In January of 2003, under the scheme "Admission of Talents", he returned to The University of Hong Kong as a Research Assistant Professor. In August of 2005, he joined the South China University of Technology as a Professor. Currently his main research interest focuses on target identification of biopharmaceuticals and nutraceuticals with omics technologies.

CONTENTS

EDITOR

Xuewu Zhang
College of Light Industry and Food Sciences
South China University of Technology
Guangzhou, China
Email: snow_dance@sina.com
Chapters 1, 4, 10, 11

CONTRIBUTORS

Note: Email addresses are provided for the corresponding authors of each chapter.

Michel Caron
Laboratoire de Biochimie des Protéines
  et Protéomique
Université Paris
Paris, France
Chapter 7

Gu Chen
College of Light Industry
  and Food Sciences
South China University of Technology
Guangzhou, China
Chapters 1, 4, 11

Je-Yoel Cho
Department of Biochemistry
Kyungpook National University
Daegu, South Korea
Email: jeycho@knu.ac.kr
Chapter 8

Yuan Gao
National Center for Toxicological
  Research
U.S. Food and Drug Administration
Jefferson, Arkansas, USA
Chapter 6

Feng Ge
Institute of Life and Health Engineering
Jinan University
Guangzhou, China
Chapter 3

Julie Hardouin
Laboratoire Polymères, Biopolymères,
  Surfaces
UMR CNRS 6270
Équipe Biofilms, Résistance, Interactions,
  Cellules-Surface
Université de Rouen
Mont-Saint-Aignan, France
Email: julie.hardouin@univ-rouen.fr
Chapter 7

Qing-Yu He
Institute of Life and Health Engineering
Jinan University
Guangzhou, China
Email: tqyhe@jnu.edu.cn
Chapter 3

Byung-Gyu Kim
Department of Biochemistry
Kyungpook National University
Daegu, South Korea
Chapter 8

Jiaying Lin
Guangdong Lung cancer Institute
Guangdong Academy of Medical Sciences
Guangzhou, China
Email: gzlinjiaying3@yahoo.com.cn
Chapter 2

Xiaoqiong Ma
Department of Nutrition and Toxicology
Zhejiang University School
   of Public Health
Hangzhou, Zhejiang, China
Chapter 9

Donna L. Mendrick
National Center for Toxicological
   Research
U.S. Food and Drug Administration
Jefferson, Arkansas, USA
Chapter 6

Lei Shi
College of Light Industry
   and Food Sciences
South China University of Technology
Guangzhou, China
Email: leishi@scut.edu.cn
Chapters 5, 11

Kaijun Xiao
College of Light Industry
   and Food Sciences
South China University of Technology
Guangzhou, China
Chapter 4

Jun Yang
Department of Nutrition and Toxicology
Zhejiang University School
   of Public Health
Hangzhou, Zhejiang, China
Email: gastate@zju.edu.cn
Chapter 9

Shang-Tian Yang
Department of Chemical
   and Biomolecular Engineering
The Ohio State University
Columbus, Ohio, USA
Email: yang.15@osu.edu
Chapter 1

Yee Leng Yap
Davos Life Science Pte. Ltd.
Singapore
Email: daniel.yap@davoslife.com
Chapters 10, 11

Li-Rong Yu
National Center for Toxicological
   Research
U.S. Food and Drug Administration
Jefferson, Arkansas, USA
Email: lirong.yu@fda.hhs.gov
Chapter 6

PREFACE

Early diagnosis of cancer remains difficult because of the lack of specific symptoms in early disease as well as the limited understanding of etiology and oncogenesis. The powerful 'omics' technologies have revolutionized the field of cancer research and opened new avenues towards biomarker discovery, identification of signaling molecules associated with cell growth, cell death, cellular metabolism and early detection of cancer.

This book attempts to provide a comprehensive overview of technologies, potential clinical applications and challenges of omics in cancer biomarker discovery. In total 11 chapters are included. Chapter 1 introduces the genomic technologies in cancer biomarker discovery, such as single nucleotide polymorphism (SNP) array, next generation sequencing and genome-wide association studies (GWAS). Chapter 2, "Transcriptomics in Cancer Biomarker Discovery", focuses on how to apply DNA microarrays in cancer biomarker discovery and how these biomarkers have impacted cancer diagnosis, treatment, and prognosis. Chapter 3 covers the advances of proteomics in cancer biomarker discovery, the challenges ahead and perspectives of proteomics for biomarker identification are also addressed. Chapter 4 discusses the current applications and future challenges of metabonomics in cancer biomarker discovery. Chapter 5 deals with peptidomics in cancer biomarker discovery. Chapter 6, "Phosphoproteomics in Cancer Biomarker Discovery", describes some examples of phosphoproteins serving as cancer biomarkers, and the applications of global and targeted quantitative phosphoproteome analysis for cancer biomarker validation and qualification. Chapter 7 demonstrates the usefulness of immunomics in cancer biomarker discovery, the top-down SERological Proteome Analysis (SERPA) and the bottom-up MAPPing (Multiple Affinity Protein Profiling) strategies are involved. Chapter 8 highlights the application of the glycoproteome analysis to the discovery of diagnostic or prognostic biomarkers for more accurate cancer diagnosis and therapy and for better understanding of cancer progression. Chapter 9 pays attention to the development of lipidomics techniques and their important roles in tumor diagnosis and therapy. Chapter 10 presents a review of bioinformatics literatures identifying cancer biomarkers from various platforms. Finally, Chapter 11 summarizes the integrative use of omics technologies for cancer biomarker discovery and the future prospects are also discussed.

The intent of this book is to bring together current advances and comprehensive reviews of omics technologies from a panel of experts from around the globe and to stimulate the development of new approaches to cancer research. The book serves as an introduction to novices in the field and a valuable reference for those already involved.

Xuewu Zhang, PhD

Chapter 1

Genomics in Cancer Biomarker Discovery

Gu Chen, Xuewu Zhang and Shang-Tian Yang*

Abstract

There is no doubt that we are approaching a time when the use of proper biomarkers will help detect, monitor and manage progression of cancer, as well as assist in therapeutic decisions. Discovery and validation of novel cancer biomarkers remain crucial goals of future patient care. Advanced genomic technologies, such as SNP array and next generation sequencing, help shape the landscapes of cancer genome and epigenome. Genome-wide association studies (GWAS), a powerful approach to identify common, low-penetrance disease loci, have been conducted in several types of cancer and have identified many novel associated loci, confirming that susceptibility to these diseases is polygenic. Though the creation of risk profiles from combinations of susceptible SNPs are not yet clinically applicable, future, large-scale GWAS holds great promise for individualized cancer screening and prevention. Epigenomic biomarkers like DNA methylation have emerged as highly promising biomarkers and are actively studied in multiple cancers. Validated as being associated with cancer risk or drug response, some DNA methylation biomarkers are being transferred into clinical use. Discovery of the genes and pathways mutated in human cancer, especially through the large-scale genome-wide sequencing, has provided key insights into the mechanisms underlying tumorigenesis and has suggested new candidate biomarkers for diagnosis, clinical intervention as well as prognosis. The comprehensive landscapes of cancer genome point out the convergence of mutations onto pathways that govern the course of tumorigenesis and indicate that rather than seeking genomics and epigenomics alterations of specific mutated genes, the combination with dynamic transcriptomics, proteomics and metabonomics of the downstream mediators or key nodal points may be preferable for future cancer biomarker discovery.

Introduction

Human cancers are primarily genetic diseases caused by genome alterations: DNA sequence changes, copy number aberrations, chromosomal rearrangements and epigenetic modifications such as DNA methylation; together they drive the development and progression of malignant transformation.[1]

It is known that cancer is caused by mutations in three types of genes and their regulatory regions: oncogenes, tumor-suppressor genes and stability genes.[2] Under mutated, oncogenes become constitutively active or active under conditions in which the normal gene is not. The unusually high activity of an oncogene, even in one allele only, will then confer a selective growth advantage to the cell and result in aberrant growth. Mutations in oncogene *KIT* genes are found

*Corresponding Author: Shang-Tian Yang—College of Light Industry and Food Sciences, South China University of Technology, Guangzhou, China. Email: yang.15@osu.edu.

Omics Technologies in Cancer Biomarker Discovery, edited by Xuewu Zhang.

related to familial gastrointestinal stromal tumors (GIST)[3,4] and also play a role in leukemia.[5] Another oncogene of GIST is *PDGFRA*. Both *KIT* and *PDGFRA* genotyping is important for GIST diagnosis and assessment of sensitivity to tyrosine kinase inhibitors.[6] Overexpression of oncogene *MET* was found in hereditary papillary renal cell carcinoma and rhabdomyosarcoma.[7,8] Mutations in proto-oncogene *RET* were found associated with multiple endocrine neoplasia, such as thyroid, parathyroid and adrenal.[9-11] In the opposite way, mutations in tumor-suppressor genes reduce their inhibition activity and lead to out-of-control growth in mutated cells. Tumor suppressor gene *RB1, APC* and *NF1* have been identified as the gatekeepers to control cell growth and their mutation were found in tumors of eye, colorectal and nervous systems, respectively.[12-15] The famous tumor-suppressor gene *TP53* encodes p53 protein as a transcription factor that normally inhibits cell growth and stimulates cell death when induced by cellular stress.[16-18] Mutations of TP53 occur in various types of cancers.[19,20] By stimulating cell birth through the cell cycle, by inhibiting normal apoptotic processes or by angiogenesis, oncogene and tumor-suppressor gene mutations similarly increase tumor cell number dramatically and drive the neoplastic process. Different from these physiological mechanisms, mutations in stability genes confer tumorigenesis through disrupting their reparation function. Some stability genes, like the mismatch repair (MMR), nucleotide-excision repair (NER) and base-excision repair (BER) genes, are responsible for repairing subtle mistakes made during normal DNA replication or induced by exposure to mutagens,[21-28] while others control chromosomal recombination and segregation; for example, the *ATM* gene alterations resulted in childhood acute lymphoblastic leukemias.[29] Once these stability genes are inactivated, higher rate of mutations occur in other genes. And the increased mutations in oncogenes and tumor-suppressor genes usually contribute to cancers or the resistance to cancer therapy.[1,30]

Other than genetic alteration, epigenetic gene regulation is another important mechanism for the onset and maintenance of cancer.[31,32] Different from alteration in DNA sequence, epigenetic alteration include heritable covalent modifications of chromatin and DNA (for example, site-specific cytosine methylation of CpG dinucleotides) that do not change DNA sequence. The most studied components of the epigenome are DNA methylation, histone modifications, as well as chromatin structure.[33,34] From the epigenetic standpoint, global DNA hypomethylation and aberrant hypermethylation of tumor suppressor gene promoters are a well-established mechanism through which cancer cells may acquire critical features on their path to transformation.[31,35]

So far the majority of cancer patients are diagnosed at late stage. If these cancers are diagnosed at early stage, the 5-year survival rate exceeds 85%.[36] While lung adenocarcinoma has an average 5-yr survival rate of 15%,[37] mainly because of late-stage detection and a paucity of late-stage treatments. Therefore, highly sensitive and specific biomarkers are urgently needed to detect cancer earlier. As more and more new treatment modalities, such as chemoprevention, gene therapy and adjuvant therapies, come into development, reliable biomarkers are needed to identify who will have true clinical benefit. For example, among every 100 women diagnosed with node-negative, ER-positive breast cancer, 85 will survive without recurrence with surgery and hormonal therapy alone. Of the 15 who will recur, chemotherapy benefits four. In other words, just 4 percent of breast cancer patients both need and can benefit from cytotoxic chemotherapy.[38] So if it is not known who is more likely to response to therapy, cytotoxic chemotherapy will punish the many to benefit the few.[38] Biomarker of drug response can help address this problem.

Genomic biomarkers come in several flavors, including disease detection and classification, treatment response prediction, treatment efficacy and prognosis. In this chapter, the technological advances of genomic approach in the discovery of cancer biomarker are summarized. The cancer associated SNPs identified through genome-wide association studies, the DNA methylation biomarkers related to cancer risk and drug response and comprehensive landscapes of cancer genomes are reviewed. And their prospects, trend and challenge in both academic research and clinical application are also discussed.

Genomic Technologies in Cancer Biomarker Discovery

The initial purpose of single nucleotide polymorphism (SNP) array was to genotype multiple SNPs simultaneously, such as the allelic imbalance (AI) and loss of heterozygosity (LOH) analysis. But due to the high resolution of SNP arrays, their use in DNA copy number analysis had been explored soon after Affymetrix released its 10K array analysis system.[39,40] Now with higher and higher throughput, SNP array technology is a powerful genomic analysis tool which can concurrently identify genome-wide genotyping and DNA dosage alterations with a wide range of applications in cancer research. It is used to determine LOH, copy number changes and DNA methylation alterations of cancer cells, as well as to investigate allelic association in cancers.[41]

Conventional Sanger sequencing is money and time consuming. Sequencing of the 3 billion base pairs of human genome took 3 to 4 years using conventional Sanger sequencing machines, e.g., ABI 3730s and costed about $300 million. Recently, four main next-generation sequencing (NGS) platforms were released: Roche 454, ABI SOLiD, Illumina Solexa and Helicos.[42] These new technologies, which are up to 200 times faster and cheaper than conventional Sanger machines, have made a variety of ambitious sequencing projects feasible, as well as contributed to the genomics biomarker discovery in cancer. Eliminating the bacterial cloning step used in traditional Sanger sequencing, NGS technologies amplify single isolated DNA molecules and analyze them in a massively parallel way. Different from analyzing terminally labeled DNA strands in Sanger sequencing, NGS use pyrosequencing on hundreds of thousands or even tens of millions of single-stranded DNA molecules immobilized on a solid surface like a glass slide or on beads.[43,44] Taking 454 as an example, single DNA strands attached to beads are amplified by separate PCR so that there is resulting clone for each DNA strand. Mixed with DNA polymerase, the beads are then immobilized in plates containing more than 1 million wells with one bead per well. Nucleotides then flow sequentially over the wells and as each nucleotide is added to form complementary DNA strands, pyrophosphate is released and detected as a chemiluminescent flash.[44] Compared with traditional Sanger sequencing machines, represented by ABI 3730, which can read about 800 bases of 100 DNA molecules simultaneously, the NGS platforms read many more DNAs in parallel but have shorter read lengths. The 454's GS FLX machine reads 400,000 DNAs of about 250 bases in length, while Illumina's Genome Analyzer and ABI's SOLiD platform can read tens of millions of DNAs up to about 35-50 bases in length. The 454 sequencer, with its relatively long reads of about 250 bases, is the choice when basic genomics information is available. High throughput of Illumina's machine makes it ideal for chromatin immunoprecipitation (ChIP) sequencing to study histone modifications, when shorter reads are sufficient since the reads only need to be long enough to find a unique match in the assembled genome. NGS were also successfully applied in SNP detection[45,46] as well as in bisulfite sequencing to analyze DNA methylation.[47] But due to the difficulties in assembling their shorter reads, NGS right now are not as effective as Sanger technology in de novo sequencing. Different NGS platforms also differ in cost per base: the 454 machine sequences are about ten times cheaper than traditional Sanger technology and Illumina and ABI are 100 times cheaper and when it becomes cheap enough, sequencing is predicted to replace SNP chips.[42] And the next-next-generation sequencing technologies are in expectation. For example, nanopore sequencing measures the change in current as a single DNA molecule pass through a tiny pore. This approach could achieve long read lengths and save costs in sample preparation, including amplification, as the only step will be purifying the DNA.

Profiling the cancer epigenome has the ultimate meaning of integrating the DNA sequence with the structural information, which leads the genetic information propagated in terms of gene regulation (epigenetics). But so far the genome-wide analyses of cancer epigenome have been hampered by technical limitation and cost required.[34] The recent development of NGS technologies, however, holds the promise of fast, reliable and cost-effective analyses. The main approaches employed thus far to identify altered epigenetic patterns in cancer cells are summarized in Table 1. Other than these high-throughput methods, small scale methylation status was usually assessed in clinical trial with the convenient real-time quantitative methylation-sensitive PCR (MethyLight).[48]

Table 1. *Comparison of approaches to study DNA methylation*

General Principle	Method	Technology Platform	Description	Ability	Limitation	Examples
Methylation sensitive restriction enzyme digestion	Restriction Landmark Genomic Scanning (RLGS)	Two dimensional gel electrophoresis	Genomic DNA digested with a methylation sensitive restriction enzyme, radiolabeled and further fractionated with a second nuclease and resolved on 2D gel	The first successful approach for the unbiased detection of differentially methylated DNA regions	Low resolution of 2D gel	Discriminate methylated CpG islands over many different tumors.[103] Identify the hypermethylated upstream region of *HoxA9* upon *p16INK4* loss in breast cancer.[104]
	Microarray-based integrated analysis of methylation by isochizomer (MIAMI)	DNA microarray	DNA microarray was cohybridized with DNA restriction fragments derived by the isoschizomers HpaII/MspI digestion	Detection of hypermethylation and polymorphism of the restriction sites represented in the array	Distribution bias of enzyme recognition site	Compare methylation status of 8091 gene promoters in lung cancer versus normal tissue.[105]
	Amplification of intermethylated sites (AIMS)	Ligation mediated PCR amplification and microarray	Before hybridization, DNA was digested with methylation sensitive endonuclease, followed by methylation insensitive isoschizomer, ligated with linker and amplified	High-throughput	Distribution bias of enzyme recognition site	Indentify epigenetic remodeling in colorectal cancer across an entire chromosome band.[106,107]

continued on next page

Table 1. *Continued*

General Principle	Method	Technology Platform	Description	Ability	Limitation	Examples
	HpaII tiny fragment enrichment by ligation-mediated PCR (HELP)	Ligation mediated PCR amplification and microarray	DNA was digested using methylcytosine sensitive HpaII and in parallel with MspI (resistant to DNA methylation), amplified by ligation mediated PCR and cohybridized to customized array	Specific	Relatively incomplete coverage due to the dependence on HpaII restriction sites	HELP.[108] Compare HELP, MeDIP, McrBC and CHARM.[109]
	McrBC (enzyme that predominantly cuts methylated DNA at RmC(N) 55-103RmC)	DNA microarray	DNA was digested with McrBC, cleaving half of the methylated DNA in the genome and all methylated CpG islands and differentially hybridized to array	Identify densely methylated regions at genomic level	Low resolution and location imprecision	Identify DNA methylation profile in an oligodendroglioma derived cell line LN-18.[110] Compare HELP, MeDIP, McrBC and CHARM.[109]
	Comprehensive high-throughput arrays for relative methylation (CHARM)	DNA microarray	Improved McrBC method with original tiling arrays and computational algorithms	Improved sensitivity and specificity, highly quantitative, relatively inexpensive, suitable for genome-wide analysis	Undetermined	Compare HELP, MeDIP, McrBC and CHARM.[109]

continued on next page

Table 1. Continued

General Principle	Method	Technology Platform	Description	Ability	Limitation	Examples
Bisulfite treatment (chemical conversion of unmethylated cytosine to uracil by sodium bisulfite or metabisulfite)	Bisulfite DNA sequencing	Sequencing	Direct sequencing of bisulfite treated DNA can be easily scaled up using NGS	The most unbiased and sensitive approach to detect methyl-cytosine residues. Interrogate simultaneously the epigenetic and genetic status of a large cohort	High cost in sequencing effort	Reveal DNA methylation profile of chromosomes 6, 20 and 22.[111] Compare methylation in 25 gene promoter in >40 samples of haematological tumors by 454 sequencing and identify a SNP associated with the methylation status of *LRP1B* gene promoter.[47]
	Illumina golden gate	bead arrays	Bisulfite treatment of DNA results in a SNP-like outcome (C/U). Four primer sets differentially amplify methylated and unmethylated alleles and hybridize to arrays	High-throughput method for analyzing large number of loci simultaneously	Expensive for bead arrays, preselected site	Quantitatively analyze preselected CGI and detect as little as 2.5% of methylation of given sequence.[112] Measure DNA methylation of CpG islands within the promoters of 2,305 genes in 206 glioblastomas.[1]

continued on next page

Table 1. *Continued*

General Principle	Method	Technology Platform	Description	Ability	Limitation	Examples
	Differential hybridization based analysis of deaminated DNA	Custom array	Size fractionated DNA is treated with bisulfite to converse cytosine to uracil that confers a differential hybridization affinity	High throughput	Expensive and low resolution	Screen methylation status of genes in prostate cancer.[113]
Immunopre-cipitation with antibody specific for methylated cytosine	Methyl-DNA immunoprecipita-tion (MeDIP)	Microarray	An antibody was used to specifically pull down 5-methyl-cytosines and the corresponding DNA was hybridized to microarrays	Specific and cost-effective. Good for detecting methylated CpG islands	Poorly sensitive, imprecise and relatively worse specificity outside CpG islands	Analyze chromo-some-wide methylated DNA.[114] Compare HELP, MeDIP, McrBC and CHARM.[109]

Genome-Wide Association Study (GWAS) of SNP in Cancer

SNP are the most common form of genetic variation in the genome, of the roughly 3 billion nucleotide bases in human genome, it is estimated that 10 million (1 in 300) are SNPs with a minor allele frequencies of at least 1%.[49] Cancer is caused by a complex interplay between genetic and environmental factors. Highly penetrant mutations explain only a small fraction of cancer cases and most genetic cancer risk is thought to be due to the contribution of many common sequence variants of low penetrance.[50] The common disease/common variant hypothesis seeks to understand the effects of common genetic variation (typically >5% minor allele frequency) that could contribute to polygenic disorders, such as cancer.[51] Based on this hypothesis, large-scale GWAS were conducted in sufficiently large number of cases and controls to identify loci associated with cancer risk, taking advantage of high-throughput genotyping approaches such as SNP arrays to evaluate a very large number of common SNPs. So far GWAS have been conducted in several types of cancer and have identified many novel disease loci, confirming that susceptibility to these diseases is polygenic.

Prior to the advent of GWAS, few studies of SNPs and cancer risk were reproducible.[51] While SNPs in the 8q24 locus as associated with prostate and colon cancer were replicated in several GWAS.[52-58] Subsequent studies suggest that certain variant in this region are also associated with increased breast and ovarian cancer risk.[59,60] Two SNPs on chromosome 5p15.33, rs2736098 in the telomerase reverse transcriptase protein and rs401681 in an intron of the *CLPTM1L* gene, were found associated with many cancer types such as lung cancer, urinary bladder, prostate and cervix cancer.[50] Nevertheless most of the common sequence variants that have recently been associated with cancer risk are particular to a single cancer type or at most two.

Colorectal Cancer

Other than the 8q24 locus, GWAS have identified multiple loci at which common variants modestly influence the risk of developing colorectal cancer (CRC); for example SNPs in 18q21(SMAD7), 11q23 were identified to associate with CRC.[58,61] Common genetic variants at the *CRAC1 (HMPS)* locus on chromosome 15q13.3 were found to influence CRC risk.[62] GWAS of four phases genotyping of totally 18,831 CRC cases and 18,540 controls found that in addition to the previously reported 8q24, 15q13 and 18q21 CRC risk loci, two previously unreported associations were identified: rs10795668, located at 10p14 and rs16892766, at 8q23.3, which tags a plausible causative gene, *EIF3H*.[63] A recent meta-analysis of two GWAS, comprising 13,315 individuals genotyped for 38,710 common tagging SNPs and replication testing in up to eight independent case-control series comprising 27,418 subjects identified four previously unreported CRC risk loci at 14q22.2, 16q22.1, 19q13.1 and 20p12.3.[64]

Prostate Cancer

Except for the 8q24.21 region, common sequence variants on 17q24.3, 17q12, 2p15, Xp11.22 and tumor suppressor gene *DAB2IP* were found to confer susceptibility to prostate cancer.[56,65,66] Haplotype analysis revealed significant associations of prostate cancer with two allele risk haplotypes on both 1q25 and 7p21.[67] Combined joint analysis of four independent GWAS of prostate cancer confirmed three previously reported loci, two independent SNPs at 8q24 and one in *HNF1B*. In addition, loci on chromosomes 7, 10 and 11 were highly significant as risk loci for prostate cancer.[68] Loci on chromosome 10 include *MSMB*, which encodes beta-microseminoprotein, a primary constituent of semen and a proposed prostate cancer biomarker and *CTBP2*, a gene with antiapoptotic activity; the locus on chromosome 7 is at *JAZF1*, a transcriptional repressor.[68]

Breast Cancer

Common variants in *CASP8* and *AKAP9* have been replicated to associate with breast cancer risk in independent GWAS.[69,70] Four plausible causative genes, *FGFR2*, *TNRC9*, *MAP3K1* and *LSP1* were identified as novel breast cancer susceptibility loci[71] and were confirmed for *FGFR2* and *TNRC9* in different studies.[72-74] GWAS identified two SNPs (rs4415084 and rs10941679) on 5p12 that confer risk, preferentially for estrogen receptor-positive breast cancer.[75] GWAS in Ashkenazi Jews (AJ) identified significant association between AJ breast cancer and a risk locus in 6q22.33, which contains candidate genes *ECHDC1*, encoding a protein involved in mitochondrial fatty acid oxidation and *RNF146*, encoding a ubiquitin protein ligase, both in known pathways of breast cancer pathogenesis.[74] Other than these recent reports in women of European descent, a new breast cancer susceptibility locus at 6q25.1 was identified especially significant among Chinese women.[76]

Lung Cancer

Multistage GWAS in lung cancer identified that variation in a region of 15q25.1 containing nicotinic acetylcholine receptors genes *CHRNA3* and *CHRNA5* contributes to lung cancer risk.[77] Similar results were obtained, that a susceptibility locus for lung cancer maps to nicotinic acetylcholine receptor subunit genes on 15q25.[78] A common variant in the nicotinic acetylcholine receptor gene cluster on chromosome 15q24 was found associated with nicotine dependence, the risk of lung cancer and peripheral arterial disease in populations of European descent.[79] And GWAS in case patients with familial lung cancer also identified familial aggregation of common sequence variants on 15q24-25.1.[80]

Other Cancers

GWAS were also conducted in many other cancers. For examples, GWAS in melanoma found common sequence variants on 20q11.22 (*CDC91L1*) confer susceptibility.[81] *ASIP* and *TYR* pigmentation variants were found associated with cutaneous melanoma and basal cell carcinoma.[82] Common variants on 1p36 and 1q42 were found associated with cutaneous basal cell carcinoma but not with melanoma.[83] GWAS of neuroblastoma, a rare children tumor identified increased risk of neuroblastoma with certain SNPs in the 6p22.3 region.[84] GWAS demonstrated that genetic variants in the common, low penetrance *PSCA* (prostate stem cell antigen) gene were associated with increased risk of sporadic diffuse-type gastric cancer in Japanese.[85] GWAS on urinary bladder cancer (UBC) from Iceland and Netherlands observed the strongest association with allele T of rs9642880 on chromosome 8q24, 30 kb upstream of *MYC*. Approximately 20% of individuals of European ancestry are homozygous for rs9642880(T) and their estimated risk of developing UBC is 1.49 times that of noncarriers.[86]

So far the risks conferred by these susceptibility alleles identified in GWAS are low, generally 1.3-fold or less. The combined effects may, however, be sufficiently large to be useful for risk prediction, targeted screening and prevention, particularly as more loci are identified.[87] In order to test the efficiency of a polygenic approach to breast cancer prevention and treatment, Pharoah et al examined the implications of common genetic variations concerning the risk of breast cancer. It is found that the risk profile generated by the known, common, moderate-risk alleles: *FGFR2, TNRC9, MAP3K1, LSP1, CASP8,* 2q35 and 8q24, does not provide sufficient discrimination to warrant individualized prevention.[88] However, useful risk stratification may be possible in the context of programs for disease prevention in the general population. Though the clinical use of single, common, low-penetrance genes is limited, a few susceptibility alleles may distinguish women who are at high risk for breast cancer from those who are at low risk, particularly in the context of population screening.[88] In another polygenic approach in prostate cancer, moderately associated SNPs in five chromosomal regions— three at 8q24 and one each at 17q12 and 17q24.3—plus a family history of prostate cancer have a cumulative and significant association with prostate cancer in the Swedish men.[89] Though the creation of cancer risk profiles based on common genetic variants identified in GWAS is not yet clinically applicable to individuals, great promise is held for the individualized cancer prevention.[51]

Epigenomic Biomarkers of Cancer

Epigenetic modifications including methylation have been identified as early events often preceding the appearance of tumor. And a lot of therapeutic intervention of these epigenetic changes have been investigated to find whether they can reverse the development of cancer.[34,90] Therefore DNA methylations have emerged as highly promising biomarkers and are being actively studied in multiple cancers (Table 2). Advantages of DNA methylation markers include that they are stable, common and involve a gain of signal. Also they are easily detected using PCR or array-based approaches in 'remote media', such as blood, sputum, urine and stool, making it well suited for clinical noninvasive detection (Table 2).[91]

Table 2. Examples of promoter methylation as cancer biomarker

Cancer	Gene	Diagnosis or Prognosis (Accuracy, Sensitivity and Specificity)	Media	References
Lung squamous cell carcinoma	GDNF, MTHFR, OPCML, TNFRSF25, TCF21, PAX8, PTPRN2, PITX2	Highly significant hypermethylations were identified in tumor tissue (p < 0.0001) and combination showed 95.6% sensitivity and specificity	Tumor sections	115
	OTX1, BARHL2, MEIS1, OC2, PAX6, IRX2, TFAP2A, EVX2	Methylated in 80-100% of tumors and hold promise as effective biomarkers for early detection	Tumor sections	35
Non-small cell lung cancer (NSCLC)	CDH13, CDKN2A/p16, FHIT, RARB, RASSF1A, ZMYND10 (BLU)	Methylation of any 2 loci was considered cancer positive with 73% sensitivity and 82% specificity	Plasma	116
	CDKN2A/p16, CDH13, RASSF1A, APC	Methylation is associated with early recurrence in patients of stage I NSCLC treated with curative surgery	Tumor sections	117
	CXCL12	Associated with unfavorable prognosis	Cell lines	118
	APC, CDKN2A/p16, HS3ST2 (3OST2), RASSF1A	Combination showed an AUC of 0.8 in early detection	Sputum	119
	CDKN2A/p16, RARB2	69% sensitivity and 87% specificity	Bronchial aspirates	120
	APC, CDKN2A/p16, RASSF1	53% sensitivity in 247 patients and in cases without a previous history of cancer, >99% specificity	Bronchial aspirates	121
Breast cancer	PITX2	Strongly correlated with increased risk of recurrence in node-negative, hormone receptor-positive, tamoxifen-treated breast cancer	Formalin-fixed paraffin-embedded and fresh frozen specimens	122, 123
	ID4	Indicate increased risk for tumour relapse	Primary tumor tissues	124
	EFEMP1	Prediction of primary breast cancer	Cancer tissues	125

continued on next page

Table 2. *Continued*

Cancer	Gene	Diagnosis or Prognosis (Accuracy, Sensitivity and Specificity)	Media	References
	PITX2	A potential prognostic or predictive biomarker in early breast cancer	Tumor tissue	126, 127
	SFRP2	A potential candidate screening marker helping to improve early breast cancer detection	Cell lines, primary breast carcinomas	128
Prostate cancer	*GSTP1*	Distinguish prostate cancer from benign prostatic hyperplasia with 75% sensitivity and 98% specificity in urine specimens and 88% specificity and 91% sensitivity in biopsy	Urine and biopsy	129
	PDLIM4, SVIL, PRIMAL, CSTP1, PTGS2	They display sensitivity of 94.7%, 75.4%, 47.4%, 89.5%, 87.7% and specificity of 90.5%, 75%, 54.2%, 95.8%, 90.2%, respectively. In combination, they were able to distinguish between prostate cancer and adjacent benign tissues with sensitivities and specificities of about 90% to 100%	Cancer tissue	130
Urothelial cancer of the bladder	*SOCS-1, STAT-1, BCL-2, DAPK, TIMP-3, E-cadherin; DAPK, BCL-2, H-TERT*	Methylations of first six genes were associated with tumor recurrence. In urinalysis methylation markers including *DAPK, BCL-2* and *H-TERT* give 81.1% sensitivity and 100% specificity	Tumor specimens, urine	131
Pancreatic cancer	*NPTX2*	The quantitative analysis of *NPTX2* hypermethylation shows 87% sensitivity and 80% specificity in diagnosis	ERCP-guided pancreatic duct brush cytology samples	132
Colorectal cancer (CRC)	*SFRP2*	Methylation occurs in 94.2%, 52.4%, 37.5% and 16.7% of patients with CRC, adenomas, hyperplstic polyps and ulcerative colitis, respectively. Of the 24 normal individuals, only 1 revealed methylated DNA	Stool	133

continued on next page

Table 2. *Continued*

Cancer	Gene	Diagnosis or Prognosis (Accuracy, Sensitivity and Specificity)	Media	References
	Vimentin gene	Vimentin was frequently methylated in advanced colorectal carcinoma	Primary carcinomas	134
	p16	The *p16* methylation score was significantly higher in patients with lymph node metastasis (p = 0.001) and tumor invasion to the veins (p = 0.020). The group with a high p16 methylation score showed significantly worse survival rates (p = 0.006)	Serum	135
Colon and other gastrointestinal cancers	ALX4	Methylation was more frequently found in colon cancer compared with noncancer controls (P < 0.0001). Using a cutoff of 41.4 pg/ml, sensitivity and specificity were 83.3% and 70%, respectively. Apart from colon adenomas and primary and metastatic colorectal cancers, *ALX4* is frequently methylated in adenocarcinomas of the gastrointestinal tract	Serum	136
Ovarian cancer	HOXA11	Methylation is strongly associated with the residual tumor after cytoreductive surgery and is a marker indicating poor prognosis	Specimens	137
	SFRP1, 2, 4, 5, SOX1, PAX1, LMX1A	Methylation correlated with recurrence and overall survival. Screening of tissues and serum revealed high sensitivity and specificity of 73.08 and 75%	Primary tumor tissues, serum	138

Aberrant DNA methylations of specific genes have been identified to associate with gender and cancer risk factors, indicating their different applications as biomarkers in different populations. In lung cancer, quantitative profiling of DNA methylation states of five cancer-associated genes (*CDH1, CDKN2A, GSTP1, MTHFR* and *RASSF1A*) revealed a high frequency of aberrant hypermethylation of *MTHFR, RASSF1A* and *CDKN2A* in lung tumors compared with control blood samples. And strong association was found between *MTHFR* hypermethylation in lung cancer and tobacco smoking while males showing higher levels of methylation in *RASSF1A*.[92] An example of prognosis marker in glioblastoma is the promoter methylation status of *MGMT*, a gene encoding DNA repair enzyme that removes alkyl groups from guanine residues, which is associated with glioblastoma sensitivity to alkylating agents.[93,94] And further integration of mutation, DNA methylation and clinical treatment data reveals a link between *MGMT* promoter methylation and a hypermutator phenotype consequent to mismatch repair deficiency in alkylating agents treated glioblastomas, an observation with potential clinical implications and confirming the methylation status of *MGMT* as one of the most important biomarkers for glioblastomas.[1]

Landscapes of Cancer Genomes

With the aid of powerful genomics and bioinformatics technologies, recent genome-wide analyses of cancer genome have figured out the landscapes of several solid cancer genomes.

Breast and Colorectal Cancers

Sjoblom et al from groups in Johns Hopkins University first reported the comprehensive sequencing and integrated analysis of breast and colorectal cancers.[95] Wood et al expanded this work and analyzed the sequence of 20,857 transcripts from 18,191 human genes, including the great majority of those that encode proteins, in the same cohort of 11 breast and 11 colorectal cancers.[96] A median of 76 and 84 nonsilent mutations per tumor were found in colorectal tumors (ranging from 49 to 111) and in breast cancers (varying from 38 to 193), respectively. Statistical analyses suggested that averaged 15 and 14 somatic mutations per tumor were likely to be responsible for driving the initiation, progression, or maintenance of tumor in colorectal and breast cancers, respectively. Mutations were found to enrich in the phosphatidylinositol-3-OH kinase (PI3K) pathway in both cancers and in nuclear factor kB (NF-kB) signaling in breast tumorigenesis. Additional pathways altered in colorectal cancer were related to cell adhesion, the cytoskeleton and the extracellular matrix. Based on the mutation data, the genomic landscapes of human breast and colorectal cancer were figured as: a handful of commonly mutated genes as mountains but are dominated by a much larger number of infrequently mutated genes as hills, which is predicted to be the general features of most solid tumors.[96] In colorectal cancer, the mountain genes are *TP53, KRAS, APC, PI3KCA* and *FBXW7*, while for breast cancer, *TP53*. But it is also indicated that the mutated genes in every two patients of the same cancer overlap to only a small extent which can explain the wide variations in tumor behavior and responsiveness to therapy. It is indicated that the "hills" but not the "mountains" dominate the cancer genome landscape. This view of cancer is consistent with the idea that a large number of mutations, each associated with a small fitness advantage, drive tumor progression.[97]

Pancreatic Cancer

Jones et al revealed the core signaling pathways in pancreatic cancer by global genomic analyses in 24 pancreatic cancers.[45] Among the 1562 somatic mutations detected from sequencing of 23,219 transcripts, representing 20,661 protein-coding genes, 25.5% were synonymous, 62.4% were missense, 3.8% were nonsense, 5.0% were small insertions and deletions and 3.3% were at splice sites or within the untranslated region (UTR). And 198 separate homozygous deletions and 144 focal high-copy amplifications were identified by SNP arrays. Totally, an average of 63 genetic alterations was found per tumor, most of which are point mutations. Candidate cancer genes identified include all genes previously known to play an important role in pancreatic cancer like *KRAS, TP53, CDK2NA* and *SMAD4* and also include numerous other genes of potential biological interest, like transcriptional activator *MLL3*; cadherin homologs *CDH10, PCDH15* and *PCDH18*; the acatenin *CTNNA2*; the dipeptidyl-peptidase *DPP6*; the angiogenesis inhibitor *BAI3*; the heterotrimeric guanine nucleotide—binding protein (G-protein)—coupled receptor *GPR133*; the guanylate cyclase *GUCY1A2*; the protein kinase *PRKCG*; and *Q9H5F0*, a gene of unknown function. Taking into account of all types of genetic alterations evaluated, 12 core signaling pathways and processes were identified, each genetically altered in 67 to 100% of the 24 cancers analyzed and had clear functional relevance to neoplasia based on annotations in the database. The major features of pancreatic tumorigenesis can be explained by the dysregulation of these 12 partially overlapping core pathways and processes through mutation but the pathway components that are altered in any individual tumor vary widely.

Glioblastomas

As the most common and lethal type of brain cancer, glioblastoma multiforme (GBM) drew attentions from different groups. Parsons et al sequenced 20,661 protein coding genes in 22 human GBM.[46] Among the genes analyzed, 685 genes (3.3%) contained at least one nonsilent somatic mutation, most of which were single-base substitutions (94%), whereas the others were small insertions, deletions, or duplications, with a mean of 47 mutations per tumor. A total of 147 amplifications and 134 homozygous deletions were identified in the 22 GBM. Candidate cancer genes identified include several genes with established roles in GBM, like *TP53, PTEN, CDKN2A, RB1, EGFR, NF1, PIK3CA* and *PIK3R1*, as well as a variety of genes that were not known to be altered in GBMs. Most notably, recurrent mutations in isocitrate dehydrogenase 1 (*IDH1*) were found in 12% of GBM patients and occurred in a large fraction of young patients. It is also found in most patients with secondary GBMs and was associated with an increase in overall survival, suggesting being a novel and potentially more specific marker for secondary GBM. Pathways enriched of alterations include TP53, RB1 and PI3K/PTEN pathways and specific pathways involved in transport of sodium, potassium and calcium ions; as well as nervous system—specific cellular pathways such as synaptic transmission, transmission of nerve impulses and axonal guidance. In the same year, researchers from The Cancer Genome Atlas reported their interim integrative analysis of DNA copy number, gene expression and DNA methylation aberrations in 206 glioblastomas and nucleotide sequence aberrations in 91 of the 206 glioblastomas.[1] Eight genes, including *PIK3R1, ERBB2, NF1* and *TP53*, were identified as significantly mutated genes and a highly interconnected network of aberrations was identified in three core pathways: RTK signaling, the p53 and RB tumor suppressor pathways.

Lung Adenocarcinoma

Based on a large collection of histopathologically well-classified primary tumors, Ding et al reported the characterization of genomic alterations in 188 lung adenocarcinoma.[98] DNA sequencing of 623 genes with known or potential links to cancer revealed 1013 somatic mutations, among which 26 genes were mutated at a high frequency, suggestive of a direct role in carcinogenesis. Other than the previously known lung adenocarcinoma genes—*TP53, KRAS, STK11, EGFR, CDKN2A, PTEN, NRAS, ERBB2, BRAF* and *PIK3CA*, the newly identified genes include tumour suppressor genes (*NF1, RB1, ATM, PTPRD, LRP1B* and *APC*) along with tyrosine kinase genes (ephrin receptor genes, *ERBB4, KDR, FGFR4* and *NTRK*) that may function as proto-oncogenes. SNP array and gene expression array demonstrated that many of these genes were also targeted by copy number alterations and/or gene expression changes. Additionally, the genetic alterations identified were enriched in components from the MAPK signaling, p53 signaling, Wnt signaling, cell cycle and mammalian target of rapamycin (mTOR) signaling pathways, suggesting the key pathways of this disease. Their results also demonstrate that lung adenocarcinoma is heterogeneous, with diverse combinations of mutations yet commonality in the main pathways affected by these mutations.

The cancer candidate genes and core pathways defined through these large scale sequencing and genomic analyses provide a relatively small subset of genes and components that could prove useful as biomarkers for diagnosis, therapeutics and prognosis of cancer.

Conclusion

A convergence of technological breakthroughs has taken genomic biomarker discovery to a new level (Fig. 1). Benefited from advances in sequencing, evolving array designs and a more sophisticated understanding of genome architecture, simple PCR, arrays or sequencing tests based on genomics and epigenomics research promise to detect cancer earlier, stratify patients into treatment classes, identify the effective cases for therapies, monitor response to therapy and predict recurrence.

GWAS of hundreds of thousands of SNPs have led to a deluge of studies of genetic variation in cancer and have identified many novel SNPs as potential markers of cancer risk. Most of these loci found in GWAS were detected at low power, indicating that many further loci will probably be detected with larger studies. So far studies of risk profiles, combinations of SNPs that may increase cancer risk, are not yet clinically applicable. In the next decade, as the body of GWAS of cancer continues to grow, individual risk profiles based on a combination of a large number of SNP markers will be created and tested. In the near future individual genotype data of SNP may be biomarkers of cancer specific screening and/or intervention.[51]

Epigenomic biomarkers like DNA methylations have emerged as highly promising biomarkers and are actively studied in multiple cancers. Some of them have been validated in large patient

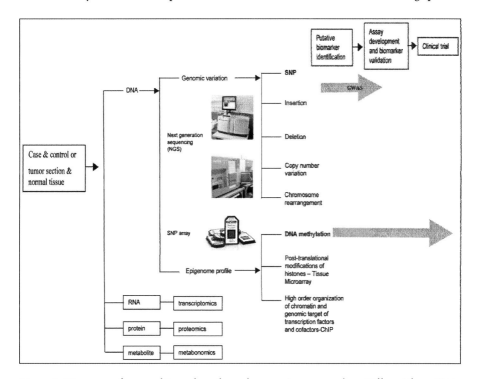

Figure 1. Discovery of cancer biomarkers through genomics approaches. Different from RNA, protein and metabolites, which are analyzed by transcriptomics, proteomics and metabonomics, respectively, DNAs from benign and cancer tissues are generally scanned by genomic approaches to identify genomic and epigenomic variation in cancer. NGS and SNP array shed light on the genome-wide analysis of cancer genome, as well as on GWAS of SNP in large cohort, though there is still a long way to go before clinical translation (indicated by the short gray arrow). Mainly based on advanced sequencing and array techniques, large scale analysis of DNA methylation have revealed a lot of promising biomarkers, some of which have been validated in large patient cohorts and were in the context of clinical trials (indicated by the long gray arrow).

cohorts and in the context of clinical trials (Table 2). Due to their clinical relevance and the ability to analyze routinely available specimens, DNA methylation biomarkers have the potential to play a leading role in forwarding molecular diagnostic tests from the bench into routine clinical use.[90]

The large scale sequencing effort of cancer genome in several solid cancers have revealed much more than anticipated, varied and complex cancer candidate genes, which can be described as: death by a thousand cuts-cancers by a thousand genes.[33] And it becomes more and more evident that the pathways, rather than individual genes, govern the course of tumorigenesis.[2,99-101] Despite the large range of mutational and epigenetic events, there is a convergence onto a finite number of pathways that drive cancer behavior. A lot of abnormalities of components of the pathways have small but additive effects on neoplasia. Tumor heterogeneity, a fundamental facet of all solid tumors,[102] can be explained by the heterogeneity among pathway components and the varied nature of mutations within individual genes. But it also represents a big challenge for cancer biomarker discovery at the genomic level.

So far genomic comparisons between cancer case and control have primarily focused on coding sequences of well annotated genes, while epigenomic analysis focused on promoter region. Recent reports showed that most of methylation alterations in colon cancer do not occur in promoters and also not in CpG islands, but in sequences up to 2 kb away, which is termed 'CpG island shores'.[32] So further investigation should include analysis of mutations in noncoding genes, mutations in noncoding regions of coding genes, relatively large deletions or insertions, amplifications, chromosomal translocations and most epigenetic changes. Such genomic analysis effort can be extended in larger cohort as well as in different types of cancers before results were verified in GWAS. And since the landscapes of cancer genome point out the convergence of mutations onto pathways that govern the course of tumorigenesis, the best hope for diagnostic and therapeutic biomarker development may lie in the discovery of physiological effects of the altered pathways and processes rather than their individual gene components. Thus, rather than seeking genomics alterations of specific mutated genes alone, with the aid of bioinformatics, integration of information from dynamic transcriptomics, proteomics and metabonomics of the downstream mediators or key nodal points may be preferable to the discovery of novel cancer biomarkers.

Acknowledgements

The authors wish to acknowledge the support of National Natural Science Foundation of China (30800609).

References

1. Chin L, Meyerson M, Aldape K et al. Comprehensive genomic characterization defines human glioblastoma genes and core pathways. Nature 2008; 455(7216):1061-1068.
2. Vogelstein B, Kinzler KW. Cancer genes and the pathways they control. Nat Med 2004; 10(8):789-799.
3. Duensing A, Heinrich MC, Fletcher CDM et al. Biology of gastrointestinal stromal tumors: KIT mutations and beyond. Cancer Invest 2004; 22(1):106-116.
4. Negri T, Pavan GM, Virdis E et al. T670X KIT Mutations in Gastrointestinal Stromal Tumors: Making Sense of Missense. J Natl Cancer Inst 2009; 101(3):194-204.
5. Sritana N, Auewarakul CU. KIT and FLT3 receptor tyrosine kinase mutations in acute myeloid leukemia with favorable cytogenetics: Two novel mutations and selective occurrence in leukemia subtypes and age groups. Exp Mol Pathol 2008; 85(3):227-231.
6. Lasota J, Miettinen M. Clinical significance of oncogenic KIT and PDGFRA mutations in gastrointestinal stromal tumours. Histopathology 2008; 53(3):245-266.
7. Takaki Y, Furihata M, Yoshikawa C et al. Sporadic bilateral papillary renal carcinoma exhibiting c-met mutation in the left kidney tumor. Journal of Urology 2000; 163(4):1241-1242.
8. Chen YY, Takita J, Mizuguchi M et al. Mutation and expression analyses of the MET and CDKN2A genes in rhabdomyosarcoma with emphasis on MET overexpression. Genes Chromosomes and Cancer 2007; 46(4):348-358.
9. Milos IN, Frank-Raue K, Wohllk N et al. Age-related neoplastic risk profiles and penetrance estimations in multiple endocrine neoplasia type 2A caused by germ line RET Cys634Trp (TGC > TGG) mutation. Endocrine-Related Cancer 2008; 15(4):1035-1041.
10. Peppa M, Boutati E, Kamakari S et al. Multiple endocrine neoplasia type 2A in two families with the familial medullary thyroid carcinoma associated G533C mutation of the RET proto-oncogene. Eur J Endocrinol 2008; 159(6):767-771.

11. Harzallah F, Barlier A, Feki M et al. Unusual presentation of multiple endocrine neoplasia type 2A in a patient with the C634R mutation of the RET-protooncogene. Ann Endocrinol (Paris) 2008; 69(6):523-525.

12. Mitter D, Rushlow D, Nowak I et al. Identification of a mutation in exon 27 of the RB1 gene associated with incomplete penetrance retinoblastoma. Fam Cancer 2009; 8(1):55-58.

13. Wood PA, Yang XM, Taber A et al. Period 2 Mutation Accelerates Apc(Min/+) Tumorigenesis. Mol Cancer Res 2008; 6(11):1786-1793.

14. Williams JP, Wu JQ, Johansson G et al. Nf1 Mutation Expands an EGFR-Dependent Peripheral Nerve Progenitor that Confers Neurofibroma Tumorigenic Potential. Cell Stem Cell 2008; 3(6):658-669.

15. Upadhyaya M, Kluwe L, Spurlock G et al. Germline and somatic NF1 gene mutation spectrum in NF1-associated malignant peripheral nerve sheath tumors (MPNSTs). Hum Mutat 2008; 29(1):74-82.

16. Vogelstein B, Lane D, Levine AJ. Surfing the p53 network. Nature 2000; 408(6810):307-310.

17. Oren M. Decision making by p53: life, death and cancer. Cell Death Differ 2003; 10(4):431-442.

18. Prives C, Hall PA. The P53 pathway. J Pathol 1999; 187(1):112-126.

19. Whibley C, Pharoah PDP, Hollstein M. p53 polymorphisms: cancer implications. Nat Rev Cancer 2009; 9(2):95-107.

20. Hrstka R, Coates PJ, Vojtesek B. Polymorphisms in p53 and the p53 pathway: roles in cancer susceptibility and response to treatment. J Cell Mol Med 2009; 13(3):440-453.

21. Modrich P. Mechanisms in eukaryotic mismatch repair. J Biol Chem 2006; 281(41):30305-30309.

22. Kolodner RD, Marsischky GT. Eukaryotic DNA mismatch repair. Curr Opin Genet Dev 1999; 9(1):89-96.

23. Waters R, Teng YM, Yu YC et al. Tilting at windmills? The nucleotide excision repair of chromosomal DNA. DNA Repair 2009; 8(2):146-152.

24. Wakasugi M, Kasashima H, Fukase Y et al. Physical and functional interaction between DDB and XPA in nucleotide excision repair. Nucleic Acids Res 2009; 37(2):516-525.

25. Tran N, Qu PP, Simpson DA et al. In Silico Construction of a Protein Interaction Landscape for Nucleotide Excision Repair. Cell Biochem Biophys 2009; 53(2):101-114.

26. Maynard S, Schurman SH, Harboe C et al. Base excision repair of oxidative DNA damage and association with cancer and aging. Carcinogenesis 2009; 30(1):2-10.

27. Kassam SN, Rainbow AJ. UV-inducible base excision repair of oxidative damaged DNA in human cells. Mutagenesis 2009; 24(1):75-83.

28. Wilson DM. Base excision repair in cancer susceptibility and neurodegeneration. Environ Mol Mutagen 2008; 49(7):531-531.

29. Pause FG, Wacker P, Maillet P et al. ATM gene alterations in childhood acute lymphoblastic leukemias (vol 21, pg 554). Hum Mutat 2003; 22(3):256-256.

30. Friedberg EC. DNA damage and repair. Nature 2003; 421(6921):436-440.

31. Jones PA, Baylin SB. The epigenomics of cancer. Cell 2007; 128(4):683-692.

32. Irizarry RA, Ladd-Acosta C, Wen B et al. The human colon cancer methylome shows similar hypo- and hypermethylation at conserved tissue-specific CpG island shores. Nat Genet 2009; 41(2):178-186.

33. Liu ET. Functional genomics of cancer. Curr Opin Genet Dev 2008; 18(3):251-256.

34. Gargiulo G, Minucci S. Epigenomic profiling of cancer cells. Int J Biochem Cell Biol 2009; 41(1):127-135.

35. Rauch TA, Zhong XY, Wu XW et al. High-resolution mapping of DNA hypermethylation and hypomethylation in lung cancer. Proc Nat Acad Sci USA 2008; 105(1):252-257.

36. Ries LAG, Reichman ME, Lewis DR et al. Cancer survival and incidence from the surveillance, epidemiology and end results (SEER) program. Oncologist 2003; 8(6):541-552.

37. Parkin DM, Bray F, Ferlay J et al. Global cancer statistics, 2002. Ca-a CA Cancer J Clin 2005; 55(2):74-108.

38. Perkel JM. Genomic biomarker discovery: bringing the genome to life. Science 2008; 319(5871):1853-1855.

39. Bignell GR, Huang J, Greshock J et al. High-resolution analysis of DNA copy number using oligonucleotide microarrays. Genome Res 2004; 14(2):287-295.

40. Zhou XF, Mok SC, Chen Z et al. Concurrent analysis of loss of heterozygosity (LOH) and copy number abnormality (CNA) for oral premalignancy progression using the Affymetrix 10K SNP mapping array. Hum Genet 2004; 115(4):327-330.

41. Mao XY, Young BD, Lu YJ. The application of single nucleotide polymorphism microarrays in cancer research. Curr Genomics 2007; 8(4):219-228.

42. von Bubnoff A. Next-generation sequencing: The race is on. Cell 2008; 132(5):721-723.

43. Dupont JM, Tost J, Jammes H et al. De novo quantitative bisulfite sequencing using the pyrosequencing technology. Anal Biochem 2004; 333(1):119-127.

44. Margulies M, Egholm M, Altman WE et al. Genome sequencing in microfabricated high-density picolitre reactors. Nature 2005; 437(7057):376-380.
45. Jones S, Zhang XS, Parsons DW et al. Core signaling pathways in human pancreatic cancers revealed by global genomic analyses. Science 2008; 321(5897):1801-1806.
46. Parsons DW, Jones S, Zhang XS et al. An integrated genomic analysis of human glioblastoma Multiforme. Science 2008; 321(5897):1807-1812.
47. Taylor KH, Kramer RS, Davis JW et al. Ultradeep bisulfite sequencing analysis of DNA methylation patterns in multiple gene promoters by 454 sequencing. Cancer Res 2007; 67(18):8511-8518.
48. Weisenberger DJ, Trinh BN, Campan M et al. DNA methylation analysis by digital bisulfite genomic sequencing and digital MethyLight. Nucleic Acids Res 2008; 36(14):4689-4698.
49. Gibbs RA, Belmont JW, Hardenbol P et al. The International HapMap Project. Nature 2003; 426(6968):789-796.
50. Rafnar T, Sulem P, Stacey SN et al. Sequence variants at the TERT-CLPTM1L locus associate with many cancer types. Nat Genet 2009; 41(2):221-227.
51. Savage SA. Cancer genetic association studies in the genome-wide age. Per Med 2008; 5(6):589-597.
52. Zanke BW, Greenwood CMT, Rangrej J et al. Genome-wide association scan identifies a colorectal cancer susceptibility locus on chromosome 8q24. Nat Genet 2007; 39(8):989-994.
53. Yeager M, Orr N, Hayes RB et al. Genome-wide association study of prostate cancer identifies a second risk locus at 8q24. Nat Genets 2007; 39(5):645-649.
54. Tomlinson I, Webb E, Carvajal-Carmona L et al. A genome-wide association scan of tag SNPs identifies a susceptibility variant for colorectal cancer at 8q24.21. Nat Genet 2007; 39(8):984-988.
55. Haiman CA, Patterson N, Freedman ML et al. Multiple regions within 8q24 independently affect risk for prostate cancer. Nat Genet 2007; 39(5):638-644.
56. Gudmundsson J, Sulem P, Manolescu A et al. Genome-wide association study identifies a second prostate cancer susceptibility variant at 8q24. Nat Genet 2007; 39(5):631-637.
57. Eeles RA, Kote-Jarai Z, Giles GG et al. Multiple newly identified loci associated with prostate cancer susceptibility. Nat Genet 2008; 40(3):316-321.
58. Tenesa A, Farrington SM, Prendergast JGD et al. Genome-wide association scan identifies a colorectal cancer susceptibility locus on 11q23 and replicates risk loci at 8q24 and 18q21. Nat Genet 2008; 40(5):631-637.
59. Schumacher FR, Feigelson HS, Cox DG et al. A common 8q24 variant in prostate and breast cancer from a large nested case-control study. Cancer Res 2007; 67(7):2951-2956.
60. Ghoussaini M, Song HL, Koessler T et al. Multiple loci with different cancer specificities within the 8q24 gene desert. J Natl Cancer Inst 2008; 100(13):962-966.
61. Broderick P, Carvajal-Carmona L, Pittman AM et al. A genome-wide association study shows that common alleles of SMAD7 influence colorectal cancer risk. Nat Genet 2007; 39(11):1315-1317.
62. Jaeger E, Webb E, Howarth K et al. Common genetic variants at the CRAC1 (HMPS) locus on chromosome 15q13.3 influence colorectal cancer risk. Nat Genet 2008; 40(1):26-28.
63. Tomlinson IPM, Webb E, Carvajal-Carmona L et al. A genome-wide association study identifies colorectal cancer susceptibility loci on chromosomes 10p14 and 8q23.3. Nat Genet 2008; 40(5):623-630.
64. Houlston RS, Webb E, Broderick P et al. Meta-analysis of genome-wide association data identifies four new susceptibility loci for colorectal cancer. Nat Genet 2008; 40(12):1426-1435.
65. Gudmundsson J, Sulem P, Rafnar T et al. Common sequence variants on 2p15 and Xp11.22 confer susceptibility to prostate cancer. Nat Genet 2008; 40(3):281-283.
66. Duggan D, Zheng SL, Knowlton M et al. Two genome-wide association studies of aggressive prostate cancer implicate putative prostate tumor suppressor gene DAB21P. J Natl Cancer Inst 2007; 99(24):1836-1844.
67. Nam RK, Zhang WW, Loblaw DA et al. A genome-wide association screen identifies regions on chromosomes 1q25 and 7p21 as risk loci for sporadic prostate cancer. Prostate Cancer Prostatic Dis 2008; 11(3):241-246.
68. Thomas G, Jacobs KB, Yeager M et al. Multiple loci identified in a genome-wide association study of prostate cancer. Nat Genet 2008; 40(3):310-315.
69. Cox A, Dunning AM, Garcia-Closas M et al. A common coding variant in CASP8 is associated with breast cancer risk. Nat Genet 2007; 39(3):352-358.
70. Frank B, Wiestler M, Kropp S et al. Association of a common AKAP9 variant with breast cancer risk: A collaborative analysis. J Natl Cancer Inst 2008; 100(6):437-442.
71. Easton DF, Pooley KA, Dunning AM et al. Genome-wide association study identifies novel breast cancer susceptibility loci. Nature 2007; 447(7148):1087-1093.
72. Stacey SN, Manolescu A, Sulem P et al. Common variants on chromosomes 2q35 and 16q12 confer susceptibility to estrogen receptor-positive breast cancer. Nat Genet 2007; 39(7):865-869.

73. Hunter DJ, Kraft P, Jacobs KB et al. A genome-wide association study identifies alleles in FGFR2 associated with risk of sporadic postmenopausal breast cancer. Nat Genet 2007; 39(7):870-874.

74. Gold B, Kirchhoff T, Stefanov S et al. Genome-wide association study provides evidence for a breast cancer risk locus at 6q22-33. Proc Natl Acad Sci USA 2008; 105(11):4340-4345.

75. Stacey SN, Manolescu A, Sulem P et al. Common variants on chromosome 5p12 confer susceptibility to estrogen receptor-positive breast cancer. Nat Genets 2008; 40(6):703-706.

76. Zheng W, Long JR, Gao YT et al. Genome-wide association study identifies a new breast cancer susceptibility locus at 6q25.1. Nat Genet 2009; 41(3):324-328.

77. Amos CI, Wu XF, Broderick P et al. Genome-wide association scan of tag SNPs identifies a susceptibility locus for lung cancer at 15q25.1. Nat Genet 2008; 40(5):616-622.

78. Hung RJ, Mckay JD, Gaborieau V et al. A susceptibility locus for lung cancer maps to nicotinic acetylcholine receptor subunit genes on 15q25. Nature 2008; 452(7187):633-637.

79. Thorgeirsson TE, Geller F, Sulem P et al. A variant associated with nicotine dependence, lung cancer and peripheral arterial disease. Nature 2008; 452(7187):638-U639.

80. Liu PY, Vikis HG, Wang DL et al. Familial aggregation of common sequence variants on 15q24-25.1 in lung cancer. J Natl Cancer Inst 2008; 100(18):1326-1330.

81. Brown KM, MacGregor S, Montgomery GW et al. Common sequence variants on 20q11.22 confer melanoma susceptibility. Nat Genet 2008; 40(7):838-840.

82. Gudbjartsson DF, Sulem P, Stacey SN et al. ASIP and TYR pigmentation variants associate with cutaneous melanoma and basal cell carcinoma. Nat Genet 2008; 40(7):886-891.

83. Stacey SN, Gudbjartsson DF, Sulem P et al. Common variants on 1p36 and 1q42 are associated with cutaneous basal cell carcinoma but not with melanoma or pigmentation traits. Nat Genet 2008; 40(11):1313-1318.

84. Maris JM, Mosse YP, Bradfield JP et al. Chromosome 6p22 locus associated with clinically aggressive neuroblastoma. N Engl J Med 2008; 358(24):2585-2593.

85. Sakamoto H, Yoshimura K, Saeki N et al. Genetic variation in PSCA is associated with susceptibility to diffuse-type gastric cancer. Nat Genet 2008; 40(6):730-740.

86. Kiemeney LA, Thorlacius S, Sulem P et al. Sequence variant on 8q24 confers susceptibility to urinary bladder cancer. Nat Genet 2008; 40(11):1307-1312.

87. Easton DF, Eeles RA. Genome-wide association studies in cancer. Hum Mol Genet 2008; 17:R109-R115.

88. Pharoah PDP, Antoniou AC, Easton DF et al. Polygenes, risk prediction and targeted prevention of breast cancer. N Engl J Med 2008; 358(26):2796-2803.

89. Zheng SL, Sun JL, Wiklund F et al. Cumulative association of five genetic variants with prostate cancer. N Engl J Med 2008; 358(9):910-919.

90. Lesche R, Eckhardt F. DNA methylation markers: a versatile diagnostic tool for routine clinical use. Curr Opin Mol Ther 2007; 9(3):222-230.

91. Anglim PP, Alonzo TA, Laird-Offringa IA. DNA methylation-based biomarkers for early detection of nonsmall cell lung cancer: an update. Mol Cancer 2008; 7:81.

92. Vaissiere T, Hung RJ, Zaridze D et al. Quantitative Analysis of DNA Methylation Profiles in Lung Cancer Identifies Aberrant DNA Methylation of Specific Genes and Its Association with Gender and Cancer Risk Factors. Cancer Res 2009; 69(1):243-252.

93. Hegi ME, Diserens A, Gorlia T et al. MGMT gene silencing and benefit from temozolomide in glioblastoma. N Engl J Med 2005; 352(10):997-1003.

94. Esteller M, Garcia-Foncillas J, Andion E et al. Inactivation of the DNA-repair gene MGMT and the clinical response of gliomas to alkylating agents. N Engl J Med 2000; 343(19):1350-1354.

95. Sjoblom T, Jones S, Wood LD et al. The consensus coding sequences of human breast and colorectal cancers. Science 2006; 314(5797):268-274.

96. Wood LD, Parsons DW, Jones S et al. The genomic landscapes of human breast and colorectal cancers. Science 2007; 318(5853):1108-1113.

97. Futreal PA, Coin L, Marshall M et al. A census of human cancer genes. Nat Rev Cancer 2004; 4(3):177-183.

98. Ding L, Getz G, Wheeler DA et al. Somatic mutations affect key pathways in lung adenocarcinoma. Nature 2008; 455(7216):1069-1075.

99. Jones S, Chen WD, Parmigiani G et al. Comparative lesion sequencing provides insights into tumor evolution. Proc Natl Acad Sci USA 2008; 105(11):4283-4288.

100. Chittenden TW, Howe EA, Culhane AC et al. Functional classification analysis of somatically mutated genes in human breast and colorectal cancers. Genomics 2008; 91(6):508-511.

101. Beerenwinkel N, Antal T, Dingli D et al. Genetic progression and the waiting time to cancer. Plos Computational Biology 2007; 3(11):2239-2246.

102. Owens AH, Coffey DS, Baylin SB. Tumor Cell Heterogeneity. New York: Academic Press, 1982.

103. Costello JF, Fruhwald MC, Smiraglia DJ et al. Aberrant CpG-island methylation has nonrandom and tumour-type-specific patterns. Nat Genet 2000; 24(2):132-138.
104. Reynolds PA, Sigaroudinia M, Zardo G et al. Tumor suppressor p16(INK4A) regulates polycomb-mediated DNA hypermethylation in human mammary epithelial cells. J Biol Chem 2006; 281(34):24790-24802.
105. Hatada I, Fukasawa M, Kimura M et al. Genome-wide profiling of promoter methylation in human. Oncogene 2006; 25(21):3059-3064.
106. Frigola J, Ribas M, Risques RA et al. Methylome profiling of cancer cells by amplification of inter-methylated sites (AIMS). Nucleic Acids Res 2002; 30(7).
107. Frigola J, Song J, Stirzaker C et al. Epigenetic remodeling in colorectal cancer results in coordinate gene suppression across an entire chromosome band. Nat Genet 2006; 38(5):540-549.
108. Khulan B, Thompson RF, Ye K et al. Comparative isoschizomer profiling of cytosine methylation: The HELP assay. Genome Res 2006; 16(8):1046-1055.
109. Irizarry RA, Ladd-Acosta C, Carvalho B et al. Comprehensive high-throughput arrays for relative methylation (CHARM). Genome Res 2008; 18(5):780-790.
110. Ordway JM, Bedell JA, Citek RW et al. Comprehensive DNA methylation profiling in a human cancer genome identifies novel epigenetic targets. Carcinogenesis 2006; 27(12):2409-2423.
111. Eckhardt F, Lewin J, Cortese R et al. DNA methylation profiling of human chromosomes 6, 20 and 22. Nat Genet 2006; 38(12):1378-1385.
112. Bibikova M, Lin ZW, Zhou LX et al. High-throughput DNA methylation profiling using universal bead arrays. Genome Res 2006; 16(3):383-393.
113. Yu YP, Paranjpe S, Nelson J et al. High throughput screening of methylation status of genes in prostate cancer using an oligonucleotide methylation array. Carcinogenesis 2005; 26(2):471-479.
114. Weber M, Davies JJ, Wittig D et al. Chromosome-wide and promoter-specific analyses identify sites of differential DNA methylation in normal and transformed human cells. Nat Genet 2005; 37(8):853-862.
115. Anglim PP, Galler JS, Koss MN et al. Identification of a panel of sensitive and specific DNA methylation markers for squamous cell lung cancer. Mol Cancer 2008; 7:62.
116. Hsu HS, Chen TP, Hung CH et al. Characterization of a multiple epigenetic marker panel for lung cancer detection and risk assessment in plasma. Cancer 2007; 110(9):2019-2026.
117. Brock MV, Hooker CM, Ota-Machida E et al. DNA methylation markers and early recurrence in stage I lung cancer. N Eng J Med 2008; 358(11):1118-1128.
118. Suzuki M, Mohamed S, Nakajima T et al. Aberrant methylation of CXCL12 in nonsmall cell lung cancer is associated with an unfavorable prognosis. Int J Oncol 2008; 33(1):113-119.
119. Shivapurkar N, Stastny V, Suzuki M et al. Application of a methylation gene panel by quantitative PCR for lung cancers. Cancer Lett 2007; 247(1):56-71.
120. Grote HJ, Schmiemann V, Geddert H et al. Aberrant promoter methylation of p16(INK4a), RARB2 and SEMA3B in bronchial aspirates from patients with suspected lung cancer. Int J Cancer 2005; 116(5):720-725.
121. Schmiemann V, Bocking A, Kazimirek M et al. Methylation assay for the diagnosis of lung cancer on bronchial aspirates: A cohort study. Clin Cancer Res 2005; 11(21):7728-7734.
122. Harbeck N, Nimmrich I, Hartmann A et al. Multicenter Study Using Paraffin-Embedded Tumor Tissue Testing PITX2 DNA Methylation As a Marker for Outcome Prediction in Tamoxifen-Treated, Node-Negative Breast Cancer Patients. J Clin Oncol 2008; 26(31):5036-5042.
123. Markert-Hahn C, Jaeger S, Dessauer A et al. Validity of DNA-methylation marker PITX2 to predict risk of recurrence in lymph node-negative hormone receptor-positive breast cancer patients: a transfer study. Breast Cancer Res Treat 2005; 94:S57-S57.
124. Noetzel E, Veeck J, Niederacher D et al. Promoter methylation-associated loss of ID4 expression is a marker of tumour recurrence in human breast cancer. Bmc Cancer 2008; 8:154.
125. Sadr-Nabavi A, Ramser J, Volkmann J et al. Decreased expression of angiogenesis antagonist EFEMP1 in sporadic breast cancer is caused by aberrant promoter methylation and points to an impact of EFEMP1 as molecular biomarker. Int J Cancer 2009; 124(7):1727-1735.
126. Connolly RM, Visvanathan K. PITX2 DNA methylation: a potential prognostic or predictive biomarker in early breast cancer. Pharmacogenomics 2008; 9(12):1797-1798.
127. Nimmrich I, Sieuwerts AM, Gelder MEM et al. DNA hypermethylation of PITX2 is a marker of poor prognosis in untreated lymph node-negative hormone receptor-positive breast cancer patients. Breast Cancer Res Treat 2008; 111(3):429-437.
128. Veeck J, Noetzel E, Bektas N et al. Promoter hypermethylation of the SFRP2 gene is a high-frequent alteration and tumor-specific epigenetic marker in human breast cancer. Mol Cancer 2008; 7:83.
129. Woodson K, O'Reilly KJ, Hanson JC et al. The usefulness of the detection of GSTP1 methylation in urine as a biomarker in the diagnosis of prostate cancer. J Urol 2008; 179(2):508-511.

130. Vanaja DK, Ballman KV, Morlan BW et al. PDLIM4 repression by hypermethylation as a potential biomarker for prostate cancer. Clin Cancer Res 2006; 12(4):1128-1136.
131. Friedrich MG, Toma MI, Chun JKHF et al. DNA methylation on urinalysis and as a prognostic marker in urothelial cancer of the bladder. Urologe 2007; 46(7):761-768.
132. Park JK, Ryu JK, Lee KH et al. Quantitative analysis of NPTX2 hypermethylation is a promising molecular diagnostic marker for pancreatic cancer. Pancreas 2007; 35(3):E9-E15.
133. Huang ZH, Li LH, Wang JF. Hypermethylation of SFRP2 as a potential marker for stool-based detection of colorectal cancer and precancerous lesions. Dig Dis Sci 2007; 52(9):2287-2291.
134. Shirahata A, Sakata M, Sakuraba K et al. Vimentin Methylation as a Marker for Advanced Colorectal Carcinoma. Anticancer Res 2009; 29(1):279-281.
135. Nakayama G, Hibi K, Kodera Y et al. p16 methylation in serum as a potential marker for the malignancy of colorectal carcinoma. Anticancer Res 2007; 27(5A):3367-3370.
136. Ebert MPA, Model F, Mooney S et al. Aristaless-like homeobox-4 gene methylation is a potential marker for colorectal adenocarcinomas. Gastroenterology 2006; 131(5):1418-1430.
137. Fiegl H, Windbichler G, Mueller-Holzner E et al. HOXA11 DNA methylation—A novel prognostic biomarker in ovarian cancer. Int J Cancer 2008; 123(3):725-729.
138. Su HY, Lai HC, Lin YW et al. An epigenetic marker panel for screening and prognostic prediction of ovarian cancer. Int J Cancer 2009; 124(2):387-393.

Transcriptomics in Cancer Biomarker Discovery

Jiaying Lin*

Abstract

Advances in analytical technologies have often driven revolutions in medicine. Over the last decade, microarray technology, as a powerful tool for transcriptomic analysis, has contributed enormously to our understanding of the molecular basis of cancer. Gene expression profiling offers an unparalleled opportunity to develop biomarkers that are useful in diagnosis and prognosis and in helping to achieve the goal of individualized cancer treatment. However, the limitations of the technology and the danger of inappropriate experimental processes should not be underestimated. Great challenges still exist for the routine clinical application of microarrays. This chapter will focus on how to apply DNA microarrays in cancer biomarker discovery and how these biomarkers have impacted cancer diagnosis, treatment and prognosis. We will also discuss the promises and challenges of this technology for future clinical practice.

Introduction

Cancer is a leading cause of death worldwide. According to the World Health Organization (WHO), the disease accounted for 7.9 million deaths (or around 13% of all deaths worldwide) in 2007, with an estimated 12 million deaths in 2030 (http://www.who.int/mediacentre/factsheets/fs297/en/index.html). There is an urgent need for us to better understand the molecular events of the disease and to search for more effective biomarkers for cancer prevention, early detection, drug development and personalized treatment.

A biomarker is defined as a characteristic that is objectively measured and evaluated as an indicator of normal biological processes, pathogenic processes, or pharmacological responses to therapeutic intervention.[1] Efficient biomarkers are useful not only for cancer diagnosis and for understanding pathomechanisms, but also serve as a basis for the development of therapeutics. They facilitate the combination of therapeutics with diagnostics and thus play a significant role in the development of individual medicine.

We have studied biomarkers for many years. However, the number of biomarkers that can be used in drug development and patient care is still very small. Previously, biomarkers were mainly discovered by traditional methods, such as polymerase chain reaction (PCR), immunohistochemisty (IHC), enzyme-linked immunosorbent assay (ELISA) and others. Nevertheless, these methods have many flaws; for example, they are time-consuming, offer low precision and allow only one biomarker to be screened at a time. To characterize the complete pathophysiology of cancer or to capture all of the therapeutic benefits or potential toxicities that a drug will have in various patient populations, we need methods that can screen a large number of biomarkers at a single time.[2-3] In addition, given the heterogeneous and complex nature of cancer, it is likely that many genes driving tumorigenesis

*Jiaying Lin—Medical Center of Guangdong General Hospital, Guangdong Lung Cancer Institute, Guangdong Academy of Medical Sciences, No.106, Zhongshan Road 2, Guangzhou, 510080, P.R. China. Email: gzlinjiaying3@yahoo.com.cn

Omics Technologies in Cancer Biomarker Discovery, edited by Xuewu Zhang.
©2011 Landes Bioscience.

have not yet to be identified. Traditional methods, with their limited power, are not satisfactory for this type of biomarker discovery and study.

Compared to traditional methods, transciptome analysis, which focuses on the complete set of RNA products transcribed in a given organism,[4] can provide a large-scale survey of the gene expression associated with the etiology of cancer or of the pharmacological responses to a therapeutic intervention. Among the extremely powerful techniques used in transcriptomics are DNA microarrays. DNA microarrays can simultaneously interrogate all human genes[5-7] and thus offer an ideal platform to improve biomarker discovery.[2,3,8-11]

Microarray analysis has been used since the late 1990s in the study of cancer.[12-14] Early researchers used it to compare two biological classes to identify differential gene expression. The technique was later applied to clinical cancer investigation to identify previously undiscovered subgroups,[18] predict outcomes,[19-22] reveal a metastatic signature[23] and guide the use of therapeutics.[24] Over the last decade, microarray technology has contributed to the development of sophisticated biomarkers, which are useful as predictors of cancer etiology, outcome and responsiveness to therapy.

However, the limitations of this technology and the danger of inappropriate experimental processes should not be underestimated. Challenges still exist to the routine clinical application of microarrays.[9,11,25-31]

This chapter will focus on how to apply DNA microarrays in cancer biomarker discovery and how these biomarkers have impacted cancer diagnosis, treatment and prognosis. We also discuss the promises and challenges of this technology for future clinical practice.

Current Development in Transcriptomic Technology—Focus on DNA Microarrays

Present Number of DNA Microarray Platforms and the Differences Among Them

Several transcriptomic technologies are useful for applications such as the serial analysis of gene expression,[32] massively parallel signature sequencing,[33] differential display,[34] cDNA representational difference analysis[35] and DNA microarrays.[6,12] However, DNA microarrays are by far the most successful and mature methodology for high-throughput, large-scale genomic analyses.[4]

In general, a DNA microarray utilizes thousands of probes (consisting of DNA; cDNA; or oligonucleotides, which represent specific genes) covalently attaching to a stable substrate, such as a glass slide, silicon wafer, or silica beads.[4,29] The basic principles of microarray technology are based on the hybridization of nucleic acids (i.e., two strands will always reassemble with base pairing A to T and C to G).[36]

Microarray technology evolved from Southern blotting. The first DNA array for expression profiling was described in 1987.[37] It was made by spotting cDNAs onto filter paper with a pin-spotting device. The first reported use of miniaturized microarrays for gene expression profiling was in 1995[6] and in 1997, the first complete eukaryotic genome (*Saccharomyces cerevisiae*) on a microarray was published.[38] Now, whole-genome expression profiling platforms for a large number of species are available. (http://en.wikipedia.org/wiki/DNA_microarray)

DNA arrays are classified into two main categories. In the first, DNA chips are divided into two types by the type of spotted probe they use.[10,36] One type, known as in-situ synthesis microarray, contains oligo DNA that is synthesized in situ directly on silicon slides. The other, the presynthesized microarray, contains previously synthesized cDNA or oligo DNA spotted on glass slides. The second category of DNA array is classified based on the type of hybridization.[4,27,29] A two-color array is simultaneously hybridized using two samples, each tagged with a different label (e.g., Cy3 and Cy5). In contrast, a single sample is hybridized in a one-color array; unlike in two-color systems, several different types of target and target labeling protocols exist for one-color arrays. In general, each kind of microarray has its advantages and disadvantages. For example, two-color arrays will report a relative ratio (fold change) for every gene expressed in two samples, whereas a one-color array will give an absolute value.

Currently, several commercial DNA microarray platforms are widely used, including Affymetrix, Illumina, Agilent, GE Healthcare and NCI_Operon. An overview of these platforms is given in Table 1.[93] Recently, these platforms have been evaluated as part of the MicroArray Quality Control project (MAQC).[39] The MAQC project recorded good intraplatform, as well as high interplatform, concordance, in terms of the genes that were identified as being differentially expressed in these microarray platforms. The project also recommended using fold-change ranking with a nonstringent *P* cutoff as a baseline to select genes until better, validated methods are developed and made widely available.[39]

More recent uses of DNA microarrays for transcriptomic analysis in biomedical research are not limited to the mRNA level. They are also being used to detect microRNAs[40] and alternative RNA splicing.[41] All of these applications will enable us to fully understand transcriptomes and to improve cancer biomarker discovery.

Using DNA Microarrays in Cancer Biomarker Discovery

Defining the Objectives of the Study

A microarray study is a multistep process. It begins with a well-defined biological question and the design of an experiment appropriate to that question. The principle of a microarray experimental design can be simple: compare profiles of gene expression between two or more biological sample groups and then find differentially expressed genes, which may be useful biomarkers. However, deciding which and how many sample groups should be involved in microarray experiments is not trivial, because such considerations will determine which biomarkers may be discovered and how valid the data will be.

The first step is to define the objective(s) of the study. The study objective will influence patient selection and the choice of analysis methods. There are three main objectives of microarray studies: class comparison, class prediction and class discovery.[8,25,42]

Class comparison studies focus on determining which genes are differentially expressed among the classes by using predefined classification specimens.[8,25,42] For example, if the objective is to find some interesting genes or to study physiologic and pathologic mechanisms in different types of tissue or samples under different conditions, then one may choose the class comparison study to compare the expression profiles of these samples. Examples of class comparison studies include studying the tumor biology of BRCA1-mutated tumors and sporadic tumors,[44] investigating molecular mechanisms between cells before and after experimental intervention and analyzing the intrinsic differences among mesenchymal stem cells derived from distinct origins.[45] Briefly, class comparison studies will answer the question of what makes two or more populations different.[43]

Although class prediction studies also examine differential gene expression between groups, they differ from class comparison studies in that their aim is mainly to develop a multivariate class predictor that can accurately classify a new specimen based on its expression profile.[8,25,42] That is, can the outcome of new individuals be predicted?[43] For example, a study may compare normal tissues to tumorous tissues to find meaningful genes for diagnosing a new sample as either normal or cancerous. Other examples include prognostic prediction studies[19-22] and drug prediction studies.[24]

The last kind of objective is class discovery. Its goal is to discover new taxonomies, groupings, or clusters within a collection of samples.[8,25,42] By class discovery, a study may seek a way to classify populations more accurately. Currently, researchers have identified several previously undiscovered subgroups in lung cancer,[46] breast cancer,[47] diffuse large B-cell lymphoma[18] and others. Class discovery study is also best suited for grouping genes into subsets with similar expression patterns to elucidate pathways.[25,48] This is based on the hypothesis that these genes may have similar functions.

Before starting a microarray experiment, one may choose one objective or combine two or three of them. However, regardless of the defined objective(s), the groups of samples for study must be representative and should ultimately realize the study purpose.

Table 1. *Microarray gene expression platforms included in the MAQC main study*[93]

Manufacturer	Code	Category I	Category II	Platform	Number of Probes[a]	Probe Length (bp)	Detection
Affymetrix	AFX	In situ synthesis microarray	One-color microarray	HG-U133 plus 2.0 GeneChip®	54,675*	25	Fluorescence
Agilent	AGL	In situ synthesis microarray	Two-color microarray	Whole human genome oligo microarray, G4112A	43,931	60	Fluorescence
	AG1	In situ synthesis microarray	One-color microarray			60	Fluorescence
Applied biosystems	ABI	Pre-synthesized oligos	One-color microarray	Human genome survey microarray v2.0	32,878	60	Chimiluminescence
GE healthcare	GEH	Pre-synthesized oligos microarray	One-color microarray	CodeLink™ human whole genome	54,359	30	Fluorescence
NCI_operon	NCI	Pre-synthesized oligos microarray	Two-color microarray	Operon human oligo set v3	37,632	39-70	Fluorescence
Illumina	ILM	Pre-synthesized oligos microarray (BeadChip microarray)	One-color microarray	Human-6 BeadChip, 48K v1.0	47,293	50	Fluorescence

*Indicates the number of probesets, each of which usually contains 11 pairs of probes, half of which are perfect matches and the other half are mismatches. (Reproduced from reference 93 with kind permission of Springer Science + Business Media).

Experimental Design for Microarray Studies[49]

After determining the study group, the next step is to determine the experimental design. The three essential principles of experimental design for microarrays are randomization, replication and design of experiments.[49]

The first element to consider is randomization, including the random assignment of varieties/samples/groups and the random sampling of populations,[49,50] as well as the random running of samples (e.g., not running all of the chips of an experimental condition on one day and none on another).[49,51,52] Randomization is the physical basis for the validity of statistical inferences.

The second element to determine is replication. There are three types of replication[49] (http://en.wikipedia.org/wiki/DNA_microarray): biological sample replication (e.g., using a large enough sample size to account for inherent biological variability), technical replication (e.g., performing two or more independent microarray experiments for each sample) and gene replication (e.g., spotting each gene multiple times and at least twice for each array). In general—and especially from a statistical point of view—biological replication should be preferred over technical replication and gene replication and large sample sizes are extremely important. Methods have been developed, based upon pilot data, to extrapolate the false discovery rate (FDR)[53,54] and expected discovery rate (EDR) for different sample sizes (http://www.poweratlas.org/papers/1.pdf), as well as to plan future study sample sizes[49,55,56] (http://www.poweratlas.org).

The last element is the design of the experiment. Previous reviews have discussed this issue in great detail.[49] Briefly, three common experimental designs for two-color microarray experiments were developed: reference design, incomplete block design and loop design.[49] Each design has its advantages and disadvantages. Because one-color arrays only allow one sample per chip, this experimental design is easier and generally involves sufficient replication and randomization of the samples.

Implementation of DNA Microarray Experiments

Experiment Description

When the experimental design has been determined, the experiments may be performed. Regardless of the chosen microarray platform, the experimental and analytical steps can be simplified as sample collection, RNA extraction, cDNA/cRNA synthesis, labeling with fluorescent dye, hybridization, image acquisition and quantification, data analysis and biological interpretation, although the details of the experimental steps between one-color and two-color microarrays differ (Fig. 1).[10]

Data Analysis

Once a microarray experiment is finished, the analysis begins. Briefly, data analysis starts with background (local or global) subtraction and data normalization (by global ratio, total intensity, linear regression, curvilinear analysis, or internal controls). Subsequent data analyses include class comparison, class discovery and class prediction. Data interpretation follows these analyses. The entire process can be finished using databases and special software for microarray analysis.

There are currently numerous databases and software. Each one may have different characteristics as well as functions. Compatible array analysis software, including Significance Analysis of Microarrays (SAM),[53] BioConductor and BRB ArrayTools, are publicly accessible. They include most of the statistics, bioinformatics and visualization tools. Many annotation databases, such as Gene Ontology (GO),[58] Kyoto Encyclopedia of Genes and Genomes (KEGG), Biocarta and GenMAPP, are generally used for the biologic interpretation of array data. Commercially maintained integrative databases include MetaCore, Ingenuity Systems and GeneSpring GT. Microarray analysis tools specific to cancer are also available, such as the Oncomine database,[59] Gene Logic's BioExpress® System Oncology Suite, caArray and caWorkbench. To store microarray data, various repositories are available, including the Stanford Microarray Database (SMD),[60] European Bioinformatics Institute's ArrayExpress[61,62] and the National Cancer Institute's Gene Expression Omnibus (GEO).[63,64] While primarily for storage, these repositories do offer some limited data analysis options, such as hierarchical clustering. All of the databases and software mentioned above are listed in Table 2.[4,11,42,49,57,65,66]

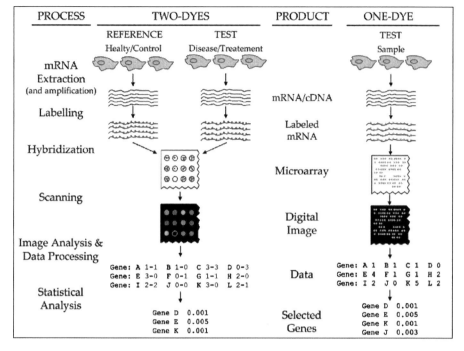

Figure 1. Schematic representation of a gene expression microarray assay. Arrows represent process (left column) and pictures or text represent the product. Differences in the protocol in one- and two-dye technologies are specific to the technology rather than to the samples or question. Reproduced from: Trevino V et al. Mol Med 2007; 13:527-541;[10] ©2007 with permission from Molecular Medicine.

Recently developed bioinformatics approaches include Module Maps, Stepwise Linkage Analysis of Microarray Signatures (SLAMS) and Cancer Outlier Profile Analysis (COPA). What these techniques have in common is the application of novel algorithms to find high-level gene expression patterns across heterogeneous microarray experiments. Large-scale initiatives, which may improve microarray data analysis and interpretation, are underway as well, such as the Cancer Genome Atlas (TCGA) project (http://cancergenome.nih.gov/index.asp) and the Cancer Biomedical Informatics Grid (caBIG™)(http://www.la-press.com/cancer-informatics-vision-cabig-a116).[66]

Overall, staggering progress has been achieved through the application of bioinformatics to cancer research, although major problems remain unsolved. Different methods may produce different results. Currently, there is no standard and ideal method for microarray data analysis and data interpretation remains a huge challenge for scientists.

The Challenges of Bioinformatics

As discussed above, the biggest challenge of bioinformatics is how to analyze and explain the data precisely. Therefore, new analytic techniques will continue to be needed to make sense of the myriad genetic changes associated with cancer, especially when those changes are subtle and poorly understood. Methodologies for data integration will be needed, as well as normalization algorithms for samples brought together from different laboratories under different conditions. We will also need the continuous development and adoption of standard formats for data storage and sample annotation. In addition, the ability to link microarray data with detailed clinical patient information, as well as the application of this data in clinical practice, need to be improved.[66]

Table 2. Website for microarray databases

Database/Software	URL
Publicly accessible softwares which cover most of the statistics, bioinformatics and visualization tools	
Significance Analysis of Microarrays (SAM)	http://www-stat.stanford.edu/~tibs/SAM/
BioConductor	http://www.bioconductor.org/
BRB ArrayTools	http://linus.nci.nih.gov/BRB-ArrayTools.html
Annotation databases with genes information, function, pathway, etc.	
LocusLink	http://www.ncbi.nlm.nih.gov/LocusLink/
Gene Ontology (GO)	http://www.ebi.ac.uk/GOA/
Kyoto Encyclopedia of Genes and Genomes (KEGG)	http://www.genome.jp/kegg/
Biocarta	http://www.biocarta.com
GenMAPP	http://www.genmapp.org
Protfun	http://www.cbs.dtu.dk/services/ProtFun
TRANSPATH	http://www.biobase.de/pages/products/transpath.html
PubMed	http://www.ncbi.nlm.nih.gov/pubmed/
Unigene	http://www.ncbi.nlm.nih.gov/sites/entrez?db = unigene
OMIM	http://www.ncbi.nlm.nih.gov/sites/entrez?db = omim
Genbank	http://www.ncbi.nlm.nih.gov/Genbank/
Genecards	http://www.genecards.org/
Structure	http://www.ncbi.nlm.nih.gov/Structure/index.shtml
UCSC	http://genome.ucsc.edu/
InterPro	http://www.ebi.ac.uk/interpro/
HPRD	http://www.hprd.org/
Swiss-Prot	http://www.expasy.org/sprot/
TRANSFAC	http://www.gene-regulation.com/pub/databases.html
Commercially maintained integrative databases and softwares	
GeneGo	http://www.genego.com/metacore.php
Ingenuity Systems	http://www.ingenuity.com
GeneSpring GT	http://www.chem.agilent.com/scripts/pds.asp?lpage = 34662
Microarray analysis tools specific to cancer	
Oncomine database	http://www.oncomine.org/main/index.jsp
Gene Logic's BioExpress® System Oncology Suite	http://www.genelogic.com/genomics/bioexpress/oncology.cfm
caArray	http://caarray.nci.nih.gov/
caWorkbench	http://amdec-bioinfo.cu-genome.org/html/caWork-Bench/newCaW3Downloads/registration.htm

continued on next page

Table 2. Continued

Database/Software	URL
Microarray data repositories	
Stanford Microarray Database (SMD)	http://genome-www5.stanford.edu/
National Cancer Institute's Gene Expression Omnibus (GEO)	http://www.ncbi.nlm.nih.gov/geo/
MAD, Jackson Labs.	http://mad.jax.org/
MGED data sharing group	http://www.mged.org/
YMD, Yale Microarray Database	http://info.mged.org/
European Bioinformatics Institute's ArrayExpress	http://www.ebi.ac.uk/arrayexpress/
Other useful database	
GoMiner	http://discover.nci.nih.gov/gominer
MatchMiner	http://discover.nci.nih.gov/matchminer/html/index.jsp
Database for Annotation, Visualization and Integrated Discovery (DAVID)	http://apps1.niaid.nih.gov/david/upload.jsp
GenePublisher	http://www.cbs.dtu.dk/services/GenePublisher

Standardization

We have outlined an entire DNA microarray experiment above. It is important to keep in mind that experimental bias exists throughout the entire process, from study design to sample collection to data generation and interpretation. This calls for standardization in platform fabrication, assay protocols and analysis methods. Until advances in microarrays occur, this standardization will maximize the acquisition of high-quality data. In addition, standardization will ease the exchange and analysis of data and ultimately bring microarrays into clinical use. Various projects and groups are working on microarray standardization, such as Minimum Information About a Microarray Experiment (MIAME)(http://www.mged.org/Workgroups/MIAME/miame.html), the MicroArray Quality Control (MAQC) project[39] and the Microarray and Gene Expression Data (MGED) group (http://www.mged.org/Workgroups/). However, more effort in microarray standardization is still needed.

Biomarker Validation and Evaluation

Once microarray data analysis is complete and potential biomarkers are identified, the next step is to confirm these biomarkers, including biomarker validation and evaluation. Presently, there are no efficient biomarker validation and evaluation platforms that can confirm data quickly and completely in a short period. Current workflows are cost- and time-consuming, need thousands of samples and the results may not be available for many years. Many studies published in papers are, in fact, incomplete validations and evaluations and therefore their data may not be applicable to clinical practice all at once.

Fortunately, this issue has been recognized and great efforts have been made to address it. The U.S. National Cancer Institute's Early Detection Research Network (EDRN) has adopted a five-phase approach to biomarker development and evaluation (see Fig. 2).[67] These phases are being successfully applied at the EDRN to decide which biomarkers are worthy of clinical validation. Currently, there are many laboratories in the EDRN, including 18 biomarker development laboratories (BDLs), 2 biomarker validation laboratories (BVLs) and 8 clinical and epidemiological centers (CECs) that are constantly monitored by the Data Management and Coordination Center, which supports the

Preclinical Exploratory	PHASE 1	Promising directions identified
Clinical Assay and Validation	PHASE 2	Clinical assay detects established disease
Retrospective Longitudinal	PHASE 3	Biomarker detects disease early before it becomes clinical and a "screen positive" rule is defined
Prospective Screening	PHASE 4	Extent and characteristics of disease detected by the test and the false referral rate are identified
Cancer Control	PHASE 5	Impact of screening on reducing the burden of disease on the population is quantifed

Figure 2. Five-phase approach to biomarker development and evaluation. Reproduced from: Pepe MS et al. J Natl Cancer Inst 2001; 93:1054-1061;[67] ©2001 with permission from Oxford University Press.

infrastructure of EDRN.[67,68] Undoubtedly these efforts will accelerate cancer biomarker research, although the platform is still not highly efficient.

Milestones of DNA Microarrays in Cancer Biomarker Discovery

Over the last decade, the introduction of microarray technology has had a profound impact on cancer research. Several important findings using microarrays have been reported and the technology is becoming more and more accessible. Microarrays have been successfully applied in the identification of drug targets,[69] in drug development[70] and in treatment validation.[71] Microarray-based gene expression profiling of human cancers has rapidly emerged as a new, powerful screening technique generating hundreds of novel diagnostic, prognostic and therapeutic targets. In Table 3, we summarize some milestone studies using microarrays in cancer research.[13,14,18,19,23,24,65,72-82]

Can DNA Microarrays Be Applied in Clinical Practice and How Far Are We So Far?

As discussed above, microarray gene expression profiling has permeated most areas of biomedical research and offered exciting opportunities for biomarker discovery and cancer diagnostics. Can microarrays, or the biomarkers generated by this technology, go further into the more demanding realm of clinical application? Some researchers remain skeptical because of the limitations of the technology.[25,26,83] They doubt that microarrays and their data are reproducible, robust and valid enough for clinical use. However, most researchers are optimistic.[8-10,27-31,43] They believe the ability of new technology that simultaneously measures many features within a single assay will provide a more accurate measurement. Because clinical variable-based models are useful but imperfect, a more precise prediction of the risk of recurrence and sensitivity to adjuvant therapy according to genomic tests will undoubtedly result in more informed decision making.[30] At least in some instances, the gene signature-based predictors were more accurate than clinical variable-based models.[84-86] Researchers concede that existing challenges will prevent the use of microarrays in routine clinical applications, but that most of these can be solved in the near future.

After summarizing a number of challenges presented in many studies,[2,3,8-11,25-31,42,43,57,66,83,87] we have classified these challenges into four questions. They are:

1. How does one obtain high-quality data?
2. How does one precisely analyze and interpret the data?
3. How does one fully validate the data in a shorter time period?
4. How does one accurately integrate the data into clinical practice?

Table 3. Milestone studies of microarray in cancer research

Year	Reports On	References
1999	Molecular classification of cancer using supervised machine learning	Golub et al[13]
2000	Molecular profiling of breast cancer	Perou et al[14]
2000	Identification of subgroups of diffuse large B-cell lymphoma with different outcomes	Alizadeh et al[18]
2002	Prediction of clinical outcomes of breast cancer	van't Veer et al[19]
2003	Identification of metastasis signature that reflected both contributions of the tumor and the host environment	Ramaswamy et al[23]
2004	Identification of prognostic profiles of adult acute myeloid leukemia	Bullinger et al[72]; Valk et al[73]
2004	Using independent samples of lymphoma to test a meta-analysis derived signature profile for predicting overall survival in diffuse large-B-cell lymphoma	Lossos et al[74]
2006	Concordance among gene-expression based predictors for breast cancer	Fan et al[75]
2006	Development of a series of oncogenic pathway signatures in human cancers	Bild et al[76]
2006	Genomic signature to guide the use of cancer chemotherapeutics	Potti et al[24]
2007	A five-gene signature and clinical outcome in nonsmall-cell lung cancer	Chen et al[77]
2007	Isolation of rare circulating tumour cells in cancer patients by microchip technology	Nagrath et al[78]
2008	Cancer proliferation gene discovery through functional genomics	Schlabach et al[79]
2008	MicroRNA expression in cytogenetically normal acute myeloid leukemia.	Marcucci et al[80]
2008	Gene expression-based survival prediction in lung adenocarcinoma: a multi-site, blinded validation study	Shedden et al[81]
2008	Epigenomics: Detailed analysis	Bonetta et al[82]

Table 3 (data from 1999 to 2006) is reproduced from reference 65, with kind permission of the Taiwan Association of Obstetrics and Gynecology (TAOG).

To solve these problems, much work must be done in many areas. The first relates to the microarray technology itself. To begin with, microarrays should be improved to make them easier to use, more precise and less expensive. For example, the whole microarray experiment could be fully automated, in which every step can be accurately controlled, so that anyone can perform the experiment correctly. We should also improve the sensitivity and specificity of the microarray for very small amounts of samples, such as biopsy samples and body fluid samples, as well as for poor quality tissues, including formalin-fixed, paraffin-embedded tissues. In addition, bioinformatics should be improved to make data analysis easier and more accurate. Futhermore, microarray experiments should be standardized, from sample collection to data generation and interpretation, facilitating the exchange of data among patients, hospitals and countries. Although several projects, such as MIAME and MAQC, are aimed at standardization, more improvement is needed.

Secondly, we need to improve our knowledge of the whole system, encompassing basic science, translation medicine, clinical medicine and so on. For example, we want to know more precisely how genes function, how pathways are regulated and how cells work. We need more accurate knowledge of the mechanisms of cancer biology, disease development and prognosis. We also need to perform studies to determine the relationship between drug effects, or toxicity and gene change. Only with more accurate knowledge can we interpret microarray data more precisely and successfully apply it to clinical practice.

In addition, it is necessary to have an efficient biomarker validation platform and workflow to confirm data quickly and completely before clinical application. Moreover, a series of clinical management procedures, based on microarray data, need to be developed, including when to use the microarray, which treatment should be chosen and how to evaluate drug effects according to the data.

The last important challenge is effective interdisciplinary communication and collaboration.[9,43,88] To achieve all of the goals mentioned above, clinicians and researchers in molecular biology, epidemiology, electronic engineering, physics, chemistry, biostatistics, computer science and mathematics, need to work together to conduct successful and efficient research into biomarker discovery and molecular diagnosis.

Although there are no current clinical applications, there is great progress in translating microarray data for clinical use. The FDA recently announced its intent to regulate molecular diagnostic tests. They released a draft guidance document covering the regulation of in vitro diagnostic multivariate index assays (IVDMIAs) (http://www.fda.gov/cdrh/oivd/guidance/1610.pdf). The FDA is also involved in the MAQC project, as discussed above, aiming to standardize the microarray platform. Two commercially available gene expression-based prognostic tests (MammaPrint for 70-gene signatures[19,22,85] and OncotypeDX for 21-gene signatures[75,86]) are currently being assessed in two large, randomized Phase III clinical trials of breast cancer. The 'Microarray In Node-negative disease may Avoid ChemoTherapy' (MINDACT) trial,[89] ongoing in Europe, is using the MammaPrint test (Agendia, Inc., Amsterdam, the Netherlands), which is a microarray-based test requiring fresh tumor samples; the result classifies breast cancer patients as having either a low or high risk of recurrence of the disease. The 'Trial Assigning Individualized Options for Treatment (RX)' (TAILORx), also called PACCT-1 trial,[90] uses OncotypeDX. OncotypeDX (Genomic Health, Redwood City, CA) is an RT–PCR-based test using formalin-fixed, paraffin-embedded tissue, to determine a recurrence score. This is a number between 0 and 100 that corresponds to a specific likelihood of breast cancer recurrence within 10 years of the initial diagnosis. In these two randomized trials, chemotherapy is compared to no chemotherapy in the population of patients defined by the gene test results. Final outcomes from these studies will not be available for many years, but the development and systematic validation of these two gene expression profiling-based tests mark a clear and exemplary translational research path for many of the aspiring novel diagnostic tests.[30]

Conclusion

Advances in analytical technologies have often driven revolutions in medicine. Over the last decade, microarray technology, as a powerful tool for transcriptomic analysis, has contributed enormously to our understanding of the molecular basis of cancer. Gene expression profiling offers an unparalleled opportunity to develop biomarkers useful in diagnosis, prognosis and to achieve the goal of individualized cancer treatment. However, the limitations of the technology and the danger of inappropriate experimental processes should not be underestimated. Prior to the routine clinical application of microarrays, we must make significant efforts to improve the microarray technology itself and our knowledge of the system and to build efficient biomarker validation platforms and accurate microarray-based clinical management procedures. Effective interdisciplinary communication and collaboration are also necessary for the successful clinical application of microarrays.

We hope that in the future DNA microarrays, or the model integrated with all "omics" technologies, may be used in clinical practice to fulfill the dream of individualized cancer care and treatment.

References

1. Biomarkers and surrogate endpoints: preferred definitions and conceptual framework. Clin Pharmacol Ther 2001; 69:89-95.
2. Hu YF, Kaplow J, He Y. From traditional biomarkers to transcriptome analysis in drug development. Curr Mol Med 2005; 5:29-38.
3. Driouch K, Landemaine T, Sin S et al. Gene arrays for diagnosis, prognosis and treatment of breast cancer metastasis. Clin Exp Metastasis 2007; 24:575-585.
4. Farber CR, Lusis AJ. Integrating global gene expression analysis and genetics. Adv Genet 2008; 60:571-601.
5. Lipshutz RJ, Morris D, Chee M et al. Using oligonucleotide probe arrays to access genetic diversity. Biotechniques 1995; 19:442-447.
6. Schena M, Shalon D, Davis RW et al. Quantitative monitoring of gene expression patterns with a complementary DNA microarray. Science 1995; 270:467-470.
7. Brown PO, Botstein D. Exploring the new world of the genome with DNA microarrays. Nat Genet 1999; 21:33-37.
8. Manning AT, Garvin JT, Shahbazi RI et al. Molecular profiling techniques and bioinformatics in cancer research. Eur J Surg Oncol 2007; 33:255-265.
9. Zhang X, Li L, Wei D et al. Moving cancer diagnostics from bench to bedside. Trends Biotechnol 2007; 25:166-173.
10. Trevino V, Falciani F, Barrera-Saldana HA. DNA microarrays: a powerful genomic tool for biomedical and clinical research. Mol Med 2007; 13:527-541.
11. Gomase VS, Tagore S, Kale KV. Microarray: an approach for current drug targets. Curr Drug Metab 2008; 9:221-231.
12. DeRisi J, Penland L, Brown PO et al. Use of a cDNA microarray to analyse gene expression patterns in human cancer. Nat Genet 1996; 14:457-460.
13. Golub TR, Slonim DK, Tamayo P et al. Molecular classification of cancer: class discovery and class prediction by gene expression monitoring. Science 1999; 286:531-537.
14. Perou CM, Sorlie T, Eisen MB et al. Molecular portraits of human breast tumours. Nature 2000; 406:747-752.
15. Welford SM, Gregg J, Chen E et al. Detection of differentially expressed genes in primary tumor tissues using representational differences analysis coupled to microarray hybridization. Nucleic Acids Res 1998; 26:3059-3065.
16. Khan J, Simon R, Bittner M et al. Gene expression profiling of alveolar rhabdomyosarcoma with cDNA microarrays. Cancer Res 1998; 58:5009-5013.
17. Agrawal D, Chen T, Irby R et al. Osteopontin identified as lead marker of colon cancer progression, using pooled sample expression profiling. J Natl Cancer Inst 2002; 94:513-521.
18. Alizadeh AA, Eisen MB, Davis RE et al. Distinct types of diffuse large B-cell lymphoma identified by gene expression profiling. Nature 2000; 403:503-511.
19. van 't Veer LJ, Dai H, van de Vijver MJ et al. Gene expression profiling predicts clinical outcome of breast cancer. Nature 2002; 415:530-536.
20. Shipp MA, Ross KN, Tamayo P et al. Diffuse large B-cell lymphoma outcome prediction by gene-expression profiling and supervised machine learning. Nat Med 2002; 8:68-74.
21. Beer DG, Kardia SL, Huang CC et al. Gene-expression profiles predict survival of patients with lung adenocarcinoma. Nat Med 2002; 8:816-824.
22. van de Vijver MJ, He YD, van't Veer LJ et al. A gene-expression signature as a predictor of survival in breast cancer. N Engl J Med 2002; 347:1999-2009.
23. Ramaswamy S, Ross KN, Lander ES et al. A molecular signature of metastasis in primary solid tumors. Nat Genet 2003; 33:49-54.
24. Potti A, Dressman HK, Bild A et al. Genomic signatures to guide the use of chemotherapeutics. Nat Med 2006; 12:1294-1300.
25. Dupuy A, Simon RM. Critical review of published microarray studies for cancer outcome and guidelines on statistical analysis and reporting. J Natl Cancer Inst 2007; 99:147-157.
26. Koscielny S. Critical review of microarray-based prognostic tests and trials in breast cancer. Curr Opin Obstet Gynecol 2008; 20:47-50.
27. Quackenbush J. Microarray analysis and tumor classification. N Engl J Med 2006; 354:2463-2472.
28. van't Veer LJ, Bernards R. Enabling personalized cancer medicine through analysis of gene-expression patterns. Nature 2008; 452:564-570.
29. Heidecker B, Hare JM. The use of transcriptomic biomarkers for personalized medicine. Heart Fail Rev 2007; 12:1-11.
30. Pusztai L. Chips to bedside: incorporation of microarray data into clinical practice. Clin Cancer Res 2006; 12:7209-7214.

31. Sotiriou C, Piccart MJ. Taking gene-expression profiling to the clinic: when will molecular signatures become relevant to patient care? Nat Rev Cancer 2007; 7:545-553.

32. Velculescu VE, Zhang L, Vogelstein B et al. Serial analysis of gene expression. Science 1995; 270:484-487.

33. Brenner S, Johnson M, Bridgham J et al. Gene expression analysis by massively parallel signature sequencing (MPSS) on microbead arrays. Nat Biotechnol 2000; 18:630-634.

34. Liang P, Pardee AB. Differential display of eukaryotic messenger RNA by means of the polymerase chain reaction. Science 1992; 257:967-971.

35. Hubank M, Schatz DG. Identifying differences in mRNA expression by representational difference analysis of cDNA. Nucleic Acids Res 1994; 22:5640-5648.

36. Nagasaki K, Miki Y. Gene expression profiling of breast cancer. Breast Cancer 2006; 13:2-7.

37. Kulesh DA, Clive DR, Zarlenga DS et al. Identification of interferon-modulated proliferation-related cDNA sequences. Proc Natl Acad Sci USA 1987; 84:8453-8457.

38. Lashkari DA, DeRisi JL, McCusker JH et al. Yeast microarrays for genome wide parallel genetic and gene expression analysis. Proc Natl Acad Sci USA 1997; 94:13057-13062.

39. Shi L, Reid LH, Jones WD et al. The MicroArray Quality Control (MAQC) project shows inter- and intraplatform reproducibility of gene expression measurements. Nat Biotechnol 2006; 24:1151-1161.

40. Schetter AJ, Leung SY, Sohn JJ et al. MicroRNA expression profiles associated with prognosis and therapeutic outcome in colon adenocarcinoma. JAMA 2008; 299:425-436.

41. Relogio A, Ben-Dov C, Baum M et al. Alternative splicing microarrays reveal functional expression of neuron-specific regulators in Hodgkin lymphoma cells. J Biol Chem 2005; 280:4779-4784.

42. Chen JJ. Key aspects of analyzing microarray gene-expression data. Pharmacogenomics 2007; 8:473-482.

43. Shi L, Perkins RG, Fang H et al. Reproducible and reliable microarray results through quality control: good laboratory proficiency and appropriate data analysis practices are essential. Curr Opin Biotechnol 2008; 19:10-18.

44. Berns EM, van Staveren IL, Verhoog L et al. Molecular profiles of BRCA1-mutated and matched sporadic breast tumours: relation with clinico-pathological features. Br J Cancer 2001; 85:538-545.

45. Tsai MS, Hwang SM, Chen KD et al. Functional network analysis of the transcriptomes of mesen-chymal stem cells derived from amniotic fluid, amniotic membrane, cord blood and bone marrow. Stem Cells 2007; 25:2511-2523.

46. Shibata T, Uryu S, Kokubu A et al. Genetic classification of lung adenocarcinoma based on array-based comparative genomic hybridization analysis: its association with clinicopathologic features. Clin Cancer Res 2005; 11:6177-185.

47. Sorlie T, Perou CM, Tibshirani R et al. Gene expression patterns of breast carcinomas distinguish tumor subclasses with clinical implications. Proc Natl Acad Sci USA 2001; 98:10869-10874.

48. Liu CC, Chen WS, Lin CC et al. Topology-based cancer classification and related pathway mining using microarray data. Nucleic Acids Res 2006; 34:4069-4080.

49. Page GP, Zakharkin SO, Kim K et al. Microarray analysis. Methods Mol Biol 2007; 404:409-430.

50. Rubin DB. Practical implications of modes of statistical inference for causal effects and the critical role of the assignment mechanism. Biometrics 1991; 47:1213-1234.

51. Kerr MK, Churchill GA. Statistical design and the analysis of gene expression microarray data. Genet Res 2001; 77:123-128.

52. Kerr MK, Churchill GA. Experimental design for gene expression microarrays. Biostatistics 2001; 2:183-201.

53. Tusher VG, Tibshirani R, Chu G. Significance analysis of microarrays applied to the ionizing radia-tion response. Proc Natl Acad Sci USA 2001; 98:5116-5121.

54. McLachlan GJ, Bean RW, Peel D. A mixture model-based approach to the clustering of microarray expression data. Bioinformatics 2002; 18:413-422.

55. Page GP, Edwards JW, Gadbury GL et al. The PowerAtlas: a power and sample size atlas for microar-ray experimental design and research. BMC Bioinformatics 2006; 7:84.

56. Jorstad TS, Midelfart H, Bones AM. A mixture model approach to sample size estimation in two-sample comparative microarray experiments. BMC Bioinformatics 2008; 9:117.

57. Lotze MT, Wang E, Marincola FM et al. Workshop on cancer biometrics: identifying biomarkers and surrogates of cancer in patients: a meeting held at the Masur Auditorium, National Institutes of Health. J Immunother 2005; 28:79-119.

58. Harris MA, Clark J, Ireland A et al. The Gene Ontology (GO) database and informatics resource. Nucleic Acids Res 2004; 32:D258-261.

59. Rhodes DR, Yu J, Shanker K et al. ONCOMINE: a cancer microarray database and integrated data-mining platform. Neoplasia 2004; 6:1-6.

60. Ball CA, Awad IA, Demeter J et al. The Stanford Microarray Database accommodates additional microarray platforms and data formats. Nucleic Acids Res 2005; 33:D580-2.
61. Brazma A, Parkinson H, Sarkans U et al. ArrayExpress—a public repository for microarray gene expression data at the EBI. Nucleic Acids Res 2003; 31:68-71.
62. Parkinson H, Sarkans U, Shojatalab M et al. ArrayExpress—a public repository for microarray gene expression data at the EBI. Nucleic Acids Res 2005; 33:D553-555.
63. Barrett T, Suzek TO, Troup DB et al. NCBI GEO: mining millions of expression profiles—database and tools. Nucleic Acids Res 2005; 33:D562-566.
64. Barrett T, Troup DB, Wilhite SE et al. NCBI GEO: mining tens of millions of expression profiles—database and tools update. Nucleic Acids Res 2007; 35:D760-765.
65. Wang TH, Chao A. Microarray analysis of gene expression of cancer to guide the use of chemotherapeutics. Taiwan J Obstet Gynecol 2007; 46:222-229.
66. Hanauer DA, Rhodes DR, Sinha-Kumar C et al. Bioinformatics approaches in the study of cancer. Curr Mol Med 2007; 7:133-141.
67. Pepe MS, Etzioni R, Feng Z et al. Phases of biomarker development for early detection of cancer. J Natl Cancer Inst 2001; 93:1054-1061.
68. Maruvada P, Srivastava S. Biomarkers for cancer diagnosis: implications for nutritional research. J Nutr 2004; 134:1640S-1645S; discussion 1664S-1666S, 1667S-167.
69. Kozian DH, Kirschbaum BJ. Comparative gene-expression analysis. Trends Biotechnol 1999; 17:73-78.
70. Gray NS, Wodicka L, Thunnissen AM et al. Exploiting chemical libraries, structure and genomics in the search for kinase inhibitors. Science 1998; 281:533-538.
71. Marton MJ, DeRisi JL, Bennett HA et al. Drug target validation and identification of secondary drug target effects using DNA microarrays. Nat Med 1998; 4:1293-1301.
72. Bullinger L, Dohner K, Bair E et al. Use of gene-expression profiling to identify prognostic subclasses in adult acute myeloid leukemia. N Engl J Med 2004; 350:1605-1616.
73. Valk PJ, Verhaak RG, Beijen MA et al. Prognostically useful gene-expression profiles in acute myeloid leukemia. N Engl J Med 2004; 350:1617-1628.
74. Lossos IS, Czerwinski DK, Alizadeh AA et al. Prediction of survival in diffuse large-B-cell lymphoma based on the expression of six genes. N Engl J Med 2004; 350:1828-1837.
75. Fan C, Oh DS, Wessels L et al. Concordance among gene-expression-based predictors for breast cancer. N Engl J Med 2006; 355:560-569.
76. Bild AH, Yao G, Chang JT et al. Oncogenic pathway signatures in human cancers as a guide to targeted therapies. Nature 2006; 439:353-357.
77. Chen HY, Yu SL, Chen CH et al. A five-gene signature and clinical outcome in nonsmall-cell lung cancer. N Engl J Med 2007; 356:11-20.
78. Nagrath S, Sequist LV, Maheswaran S et al. Isolation of rare circulating tumour cells in cancer patients by microchip technology. Nature 2007; 450:1235-1239.
79. Schlabach MR, Luo J, Solimini NL et al. Cancer proliferation gene discovery through functional genomics. Science 2008; 319:620-624.
80. Marcucci G, Radmacher MD, Maharry K et al. MicroRNA expression in cytogenetically normal acute myeloid leukemia. N Engl J Med 2008; 358:1919-1928.
81. Shedden K, Taylor JM, Enkemann SA et al. Gene expression-based survival prediction in lung adenocarcinoma: a multi-site, blinded validation study. Nat Med 2008; 14:822-827.
82. Bonetta L. Epigenomics: Detailed analysis. Nature 2008; 454:795-798.
83. Modlich O, Prisack HB, Bojar H. Breast cancer expression profiling: the impact of microarray testing on clinical decision making. Expert Opin Pharmacother 2006; 7:2069-2078.
84. Hess KR, Anderson K, Symmans WF et al. Pharmacogenomic predictor of sensitivity to preoperative chemotherapy with paclitaxel and fluorouracil, doxorubicin and cyclophosphamide in breast cancer. J Clin Oncol 2006; 24:4236-4244.
85. Buyse M, Loi S, van't Veer L et al. Validation and clinical utility of a 70-gene prognostic signature for women with node-negative breast cancer. J Natl Cancer Inst 2006; 98:1183-1192.
86. Paik S, Shak S, Tang G et al. A multigene assay to predict recurrence of tamoxifen-treated, node-negative breast cancer. N Engl J Med 2004; 351:2817-2826.
87. Srivastava S, Gopal-Srivastava R. Biomarkers in cancer screening: a public health perspective. J Nutr 2002; 132:2471S-2475S.
88. Lin DW, Nelson PS. Microarray analysis and tumor classification. N Engl J Med 2006; 355:960; author reply 960.
89. Mook S, Van't Veer LJ, Rutgers EJ et al. Individualization of therapy using Mammaprint: from development to the MINDACT Trial. Cancer Genomics Proteomics 2007; 4:147-155.
90. Sparano JA. TAILORx: trial assigning individualized options for treatment (Rx). Clin Breast Cancer 2006; 7:347-50.

91. Fehrmann RS, Li XY, van der Zee AG et al. Profiling studies in ovarian cancer: a review. Oncologist 2007; 12:960-966.
92. Sawyers CL. The cancer biomarker problem. Nature 2008; 452:548-52.
93. Leming Shi, Roger G. Perkins, Weida Tong. The Current Status of DNA Microarrays(M)//Kilian Dill, Robin Hui Liu and Piotr Grodzinski. Integrated Analytical Systems-Microarrays Preparation, Microfluidics, Detection Methods and Biological Applications. USA. Springer New York: 2009; 3-24.

CHAPTER 3

Proteomics in Cancer Biomarker Discovery

Feng Ge and Qing-Yu He*

Abstract

In recent years the discovery of cancer biomarkers has become a major focus of cancer research, which holds promise for early tumor detection and diagnosis and disease recurrence monitoring and therapeutic efficacy. Proteomics technologies are emerging as a useful tool in the discovery of cancer biomarkers: Substantial technological advances in proteomics and related computational science have been made in the past few years. These advances overcome in part the complexity and heterogeneity of the human proteome, permitting the quantitative analysis and identification of protein changes associated with tumor development. With the advent of new and improved proteomic technologies, it is possible to discover new biomarkers for the early detection and treatment of cancer. This chapter covers a selection of advances in the realm of proteomics and the discovery of cancer biomarkers in recent years. The challenges ahead and perspectives of proteomics for biomarker identification are also addressed. With a wealth of information that can be applied to a broad spectrum of biomarker research projects, this chapter serves as a reference for scientists working in proteomics and cancer.

Introduction

During the past three decades, there has been significant progress in both the understanding and treatment of cancer. However, cancer remains a major public health challenge and cause of mortality.[1] As an important biological indicator of cancer status and progression for the physiological state of the cell at a specific time, biomarkers represent powerful tools for monitoring the course of cancer and gauging the efficacy and safety of novel therapeutic agents. Biomarker, defined by the USA National Cancer Institute as 'a biological molecule found in blood, other body fluids, or tissues that is a sign of a normal or abnormal process, or of a condition or disease. A biomarker may be used to see how well the body responds to a treatment for a disease or condition.'[2] A cancer biomarker is a molecular signature that indicates the physiologic and pathologic changes in a particular tissue or cell type during cancer development. Cancer biomarkers can aid in diagnosis and/or patient management by refining staging and predicting or monitoring response to treatment.[3,4] They can have tremendous therapeutic impact on clinical oncology, especially if the biomarkers are detected before clinical symptoms or able to be real-time monitoring of drug response. There is a critical need for expedited development of biomarkers to improve diagnosis and treatment of cancer.

Cancer biomarker discovery has greatly benefited from the proteomic approach, as highlighted in a number of reviews.[5-8] Advancing technologies, particularly the evolution of 2-dimensional gel electrophoresis (2-DE) based approaches into liquid chromatography (LC) based high-resolution

*Corresponding Author: Qing-Yu He—Institute of Life and Health Engineering, Jinan University, Guangzhou 510632, China. E-mail: tqyhe@jnu.edu.cn

Omics Technologies in Cancer Biomarker Discovery, edited by Xuewu Zhang.
©2011 Landes Bioscience.

tandem mass spectrometry (MS/MS), have radically improved the speed and precision of identifying and measuring target proteins in biological fluids and other samples. The emerging technology of quantitative proteomics also provides a unique opportunity to reveal static or perturbation-induced changes in a protein profile. It is possible to discover biomarkers that are able to reliably and accurately predict outcomes during cancer treatment and management. Besides, the newer technologies provide higher analytical capabilities, employing automated liquid handling systems, fractionation techniques and bioinformatics tools for greater sensitivity and resolving power, more robust and higher throughput sample processing and greater confidence in analytical results can be obtained. Proteomics offers cutting-edge capabilities to accelerate the translation of basic discoveries into daily clinical practice. Continued progress of techniques and methods to determine the abundance and status of proteins holds great promise for the future study of cancer and the discovery of cancer biomarkers.

In this chapter, we will review recent advances in proteomics research and explore the applications of these technologies primarily in the field of cancer biomarker discovery.

Proteomic Technologies in Cancer Biomarker Discovery

Proteomics is an exciting area of research that holds promise for the future. It is widely accepted that proteomics has the potential to identify new diagnostic and prognostic biomarkers and drug targets for the development of new therapeutic approaches for fighting disease. In the last decade, many technological advances for the qualitative and quantitative characterization of proteins have been reported. With respect to gene expression profiling, proteomic analysis can give important complementary information, such as the direct readout of protein expression levels and the mapping of posttranslational modifications and protein-protein interactions. The progressive introduction of proteomic approaches in cancer research is expected to help improving the understanding of the molecular pathogenesis of cancer and to allow the discovery of novel diagnostic and prognostic markers of cancers.[9]

The use of proteomics-based approaches for biomarker research forms part of a so-called pipeline from discovery through validation, to translation and to the clinic. Samples are selected to address the clinical question, proteins are separated by use of specific complementary techniques (each of which visualizes only a small proportion of the proteome), data are analyzed to identify potential differentially expressed proteins or biomarkers, suitable assays are developed, results are validated and ultimately large-scale clinical trials are performed (Fig. 1).

Proteomic analysis is achieved by a combination of techniques that are designed to profile, quantitate and identify proteins or peptides, commonly with no preconceived restrictions about the molecules involved. As shown in Figure 1, proteomic research can rely on a panel of several different technologies. To date, none of these technologies has demonstrated clear superiority over others with respect to parameters such as throughput, reproducibility and sensitivity. Most importantly, none of them has demonstrated the ability of achieving full, comprehensive proteome coverage of complex protein samples such as cell lysates, tissues or body fluids. The proteomic tools described below can be thus considered to yield complementary information.

A classical approach is 2-DE for protein separation and quantitation, with mass spectrometry to identify molecules of interest.[10] In the first dimension, proteins are separated on the basis of charge by isoelectric focusing with immobilized pH gradients; separation in the second dimension is by molecular weight by use of PAGE with sodium dodecylsulfate (SDS). Proteins are generally visualized by staining with silver, Coomassie Blue, or fluorescent dyes such as SyproRuby and up to 2000 proteins can be visualized in a standard format gel. The gel profiles of different samples can be compared by use of analysis softwares to identify sets of differentially expressed proteins. These proteins can be isolated from gels, digested into peptides with enzymes such as trypsin and the resulting peptides sequenced by use of mass spectrometry for protein identification.[11] More recently, fluorescence-based difference gel electrophoresis (DIGE) was developed to simplify analysis and improve reproducibility and throughput, which involves the addition of fluors 1 and 2 to the control and treatment samples, respectively.[12] The advantages of the 2-DE method are the powerful

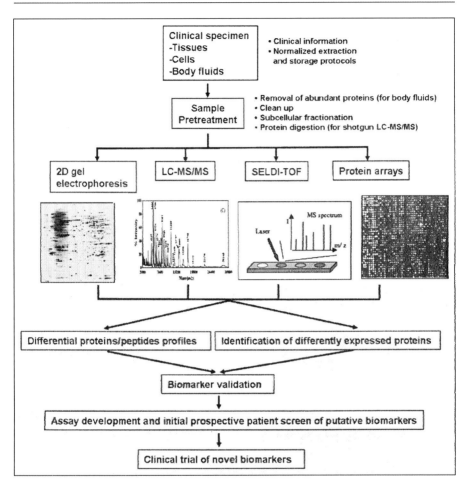

Figure 1. Schematic representation of a typical experiment oriented to the identification of biomarkers of human cancer based on proteomics approaches. Analysis of proteome can be performed by different techniques. According to the source of the sample and the designed experimental approach, samples undergo a pretreatment process, then they are analyzed by different methods: 2D gel electrophoresis (2-DE); Surface-Enhanced Laser Desorption Ionization (SELDI) coupled to Time-Of-Flight (TOF) Mass Spectrometry (MS); Single or multidimensional Liquid Chromatography coupled to tandem Mass Spectrometry (LC-MS/MS); and Protein arrays. After proteomic patterns are obtained by 2-DE or MS, bioinformatic algorithms are trained to distinguish healthy from malignant states. These discriminatory algorithms are validated by testing on independent blinded sample sets. If the algorithm proves to have sufficient sensitivity and specificity, the biomarker pattern can be applied clinically, or alternatively, putative biomarkers can be identified. Once biomarkers are identified, traditional clinical assays can be developed.

technology for protein separation, relative simultaneous quantification of proteins on gel images and identification of protein isoforms and posttranslational modifications (e.g., phosphorylation, hydroxylation, methylation, glycosylation, acetylation and oxidation). The disadvantages include limited success for the separation of proteins (e.g., membrane proteins and plasma proteins) with extremes in isoeletric points, molecular weights, abundance and hydrophobicity and inability to provide automation or absolutely quantitative information.[13]

Multidimensional liquid chromatography (MDLC) combined with tandem mass spectrometry is increasingly used in so-called shotgun strategies, where proteins in samples are digested by use of trypsin and the peptides are separated and identified (e.g., by multidimensional protein identification technology, MuDPIT).[14] Although information about the form of intact proteins might be lost, these approaches are very sensitive, enabling to identify hundreds or thousands of proteins in a sample. Quantitative comparisons can be made by use of label free strategies or tags such as stable isotope labeling with amino acids in culture (SILAC). SILAC is a simple in vivo labeling strategy for mass spectrometry-based quantitative proteomics. It relies on the metabolic incorporation of nonradioactive heavy isotopic forms of amino acids into cellular proteins, which can be readily distinguished in a mass spectrometer. As the samples are mixed before processing in the SILAC methodology, the sample handling errors are also minimized.[15,16] Alternatively, isobaric tags for relative and absolute quantitation (iTRAQ) compares up to four samples labeled with different tags that are indistinguishable in MDLC, but on cleavage in tandem mass spectrometry generate distinct signature peaks, the relative amounts of which reflect the amounts of the peptide and hence protein in every sample.[17] While the standard iTRAQ reagents afford monitoring of up to four different conditions in a single MS analysis, very recently novel reagents have become commercially available that enable multiplexed quantification of up to 8 samples.[18,19] The possibility of simultaneous quantitative analysis of more data points is particularly useful for studies of biological systems over multiple time points or in response to multiple treatment conditions—such as in concentration-response experiments.

A variant of MALDI-TOF, which has been widely used in biofluid clinical proteomics, is surface-enhanced laser desorption ionization (SELDI-TOF).[20] In this approach, clinical samples are applied on chip surfaces tailored to capture specific analyte classes via chromatographic interactions (hydrophobic, ionic, affinity). After chip washing, bound substances are analyzed by MALDI-TOF mass spectrometry. This approach has been extensively used to profile low molecular weight proteins (<20 kDa) in serum samples, in the search for diagnostic patterns of disease. SELDI-TOF has generated a lot of expectations from the clinical research world, especially because of its high-throughput capability. Nevertheless, several issues on the technology are still open.[21] Main concerns are on its sensitivity of detection, accuracy of quantification and its effective capability of generating reproducible diagnostic patterns in different laboratories. In spite of this general concern, attempts have been made to achieve a good interplatform reproducibility in cooperative research programs.[22]

Protein microarray platforms are being developed to rapidly screen protein function in a high-throughput manner. These can be categorized into two groups, forward-phase arrays and reverse-phase arrays.[23,24] For forward-phase arrays, a capture molecule (such as an antibody) for a target protein is immobilized onto a glass slide in a similar fashion as DNA microarray probes are robotically spotted with high density. The cellular lysate containing the target protein is incubated over the antibodies and the bound protein is detected using a labeled secondary antibody. With this method, many target proteins in one sample can be detected at once. In reverse-phase arrays, a complex protein mixture is immobilized onto a glass slide and probed with a specific antibody against a protein of interest. This method allows the detection of a protein of interest in a complex sample such as a tissue lysate or serum as a high-throughput assay by spotting a large number of samples on a single glass slide. Because of specificity of antibody binding and use of signal amplification by sensitive labeling methods, the detectable range of protein concentration can be as low as less than 10 cell equivalent.[24] Also, recently the number of commercially available antibodies, including antibodies to specific protein modifications such as phosphorylation, has grown exponentially, making the application of the protein microarrays feasible for clinical studies. Important drawbacks of protein/antibody arrays are the difficulty of establishing general binding conditions for the proteins under study and the possible interference from macromolecular complexes, not disrupted in the nondenaturing conditions employed.[25] New protein microarray platforms such as self assembling arrays are emerging, which promise a much easier and wider use of the technology to probe protein interaction and function.[26]

Complementary in terms of coverage, these proteomic technological methods all have inherent strengths and weaknesses. For instance, the combination of data from three methodologically different extensive laboratory studies with proteins already described in a published work identified 1175 non redundant proteins, only 46 of which were found in all data sets and only 195 of which were from more than one source.[27] Different methods need to be accompanied by prefractionation strategies to allow the study of proteins with low abundance or in specific subgroups.

Proteomic Biomarkers

Biomarkers for Early Detection of Cancer

It is well known that early detection and diagnosis of cancer can improve the clinical outcome more than any therapeutic intervention available for advanced stage cancers. The early detection of cancer has a potential to dramatically reduce mortality. Proteomics technology has been applied to analyses of tissue, serum, saliva and nipple aspirate fluid in an attempt to identify biomarkers of cancer and to develop screening tools with high sensitivity and specificity.

By analyzing serum using MALDI-TOF and SELDI-TOF MS, the presence of head and neck cancer could be detected with a sensitivity and specificity of 68-83.3% and 73-90%, respectively.[28-30] In prostate cancer, by using SELDI-TOF MS combined with pattern recognition based bioinformatics tool, discriminatory spectrum proteomic profiles were generated, helping discriminating men with prostate cancer from those with benign prostate.[31,32] Wang et al took an innovative approach to the early detection of prostate cancer by using autoantibodies present in serum against peptides in prostate tumor tissue and termed this emerging technology as "cancer immunomics".[33] By comparing sera from 119 prostate cancer and 138 control groups, this method could identify prostate cancer with 88.2% specificity and 81.6% sensitivity.

Proteomic analysis successfully identified 154 potential serum markers for pancreatic cancer.[34] Of these, fibrinogen γ, a protein associated with the hypercoagulable state of pancreatic cancer, discriminated cancer from normal sera. Fibrinogen γ was subsequently confirmed to be over-expressed in pancreatic cancer sera by enzymatic analysis and in tissue by immunohistochemistry relative to normal pancreas, thus it is a potential tumor marker in pancreatic cancer.[34]

In lung cancer, Maciel et al investigated whether serum proteins could discriminate lung adenocarcinoma patients from healthy donors. Results of 2-DE/MALDI-TOF showed 5 up-regulated proteins (immunoglobulin lambda chain, transthyretin monomer, haptoglobin-alfa 2 and 2 isoforms of serum amyloid protein) and 1 downregulated protein (fragment of apolipoprotein A-I) in lung adenocarcinoma patients.[35] Using 2-DE based analysis of sera from patients including 64 newly diagnosed lung cancer and 71 noncancer controls, Hanash et al identified the neuro-specific polypeptide PGP (protein gene product) 9.5 as a tumor antigen that induces a humoral immune response in lung cancer and thus may have utility in screening and diagnosis.[36] Xiao et al have utilized the SELDI technology to identify novel serum biomarkers in lung cancer for early detection.[37] In this study, serum samples from 30 lung cancer patients and 51 age- and sex-matched healthy individuals were analyzed by SELDI. Protein peaks clustering and classification analyses were performed. The constructed model then was used to test an independent set of masked serum samples from 15 lung cancer patients and 31 healthy individuals, yielding a sensitivity of 93.3% and a specificity of 96.7%.

For ovarian cancer, Ahmed et al found that changes in serum expression of haptoglobin correlated with the change of CA-125 levels before and after chemotherapy.[38] Of great significance, the technique worked well on patients with early stage disease, offering the prospect of earlier diagnosis and greatly enhancing the chance of successful treatment outcome. This has led to the development of a commercial test, termed OvaCheck, for diagnosis of ovarian cancer.

The enhanced expression of ceruloplasmin in the nasopharyngeal carcinoma patients' sera was detected by 2-DE and further confirmed by competitive enzyme-linked immunosorbent assay (ELISA).[39] These data provided potential biomarkers for early diagnosis of nasopharyngeal carcinoma.

2-D DIGE combined with nano flow liquid chromatography tandem MS was employed to investigate differentially expressed proteins in hepatocellular carcinoma (HCC). 14-3-3γ protein was found to be up-regulated in HCC.[40] 14-3-3 isoforms has been linked to carcinogenesis because they are involved in various cellular processes such as cell cycle regulation, apoptosis, proliferation and differentiation.[41] SELDI-TOF MS was also performed to identify differentially expressed proteins in HCC serum using weak cation exchange proteinchips. Protein characterization was performed by 2-DE separation and nano flow LC-MS/MS. Complement C3a was detected with differential expression in patients with chronic hepatitis C and hepatitis C virus-related HCC. This result was further validated by PS20 chip immunoassay and Western blotting.[42] Besides, the use of protein chip technology in combination with tissue microdissection has identified ferritin light subunit, adenylate kinase 3 alpha-like 1 and biliverdin reductase B in HCC.[43]

To identify proteins with colorectal cancer (CRC)-specific regulation, comparative 2-DE of individual matched normal and neoplastic colorectal tissue specimens was performed. Endocrine cell-expressed protein secretagogin exhibited a marked down-regulation in CRC tissues. This finding may represent the basis for the clinical application of secretagogin as a biomarker for a distinct subgroup of CRCs.[44] Using DIGE, a number of proteins with altered expression between primary esophageal cancer and adjacent noncancer tissues have been identified. Among them, periplakin was significantly down-regulated in esophageal cancer, which was confirmed by immunoblotting and immunohistochemistry. These results suggested that periplakin could be a useful marker for the detection of early esophageal cancer and the evaluation of tumor progression.[45]

Biomarkers for Therapeutic Targets

Various organ specific tumor and normal tissues have been analyzed to search for differentially expressed proteins, as well as protein modifications such as phosphorylation of signaling receptors, which can potentially be developed as therapeutic targets. Proteomics can be used to identify a candidate protein in preclinical studies and to confirm a candidate protein in clinical specimens. Many cancers are characterized by alternations in certain signaling pathways and identification of the aberrant pathway in a particular patient allows for targeted therapy to the specific pathway. For example, epithelial ovarian cancer is often characterized by activation of EGFR signaling pathway and targeted therapies including recombinant humanized monoclonal antibodies (rhuMAbs), such as cetuximab[46] and small molecule inhibitors such as gefitinib[47] are either in clinical use or under clinical trial for different stages of cancer. Similarly, the c-Kit and platelet derived growth factor receptor inhibitor, Imatinib, has shown remarkable success in chronic myeloid leukemia.[48]

Using 2D-DIGE and protein arrays, IKK-beta was identified as a novel endogenous marker of tumor hypoxia and as a potential therapeutic target.[49] By using global phosphoproteomic approaches based on immunoaffinity purification of phosphopeptides, activated protein kinases and their phosphorylated substrates could be identified in a preclinical study.[50]

Proteomics-based studies have widened our knowledge of transforming growth factor-β-dependent regulation of cell proliferation, apoptosis, DNA damage repair and transcription. This leads to better understanding of the transforming growth factor-β role in human breast tumorigenesis and opens the avenue for the development of novel anticancer treatments and drugs, with some of the drugs already entering clinics.[51]

The identification of antigens expressed by prostate tissue and/or prostate cancer that are recognized by T-cells or antibodies creates opportunities to develop novel immunotherapeutic approaches including tumor vaccines. Proteins expressed in prostate cancer including PSA, prostatic acid phosphatase and prostate membrane antigen have been used as immunologic targets for immunotherapy.[52]

Biomarkers for Prediction of Recurrence and Survival

Prognostication and the variability of tumor responses to radio-/chemo-therapeutic agents are a topic of major interest in current cancer research. The advances in proteomic research will lead to a plethora of new molecular markers, which are likely to be correlated with disease activity, progression and survival. Rapid developments in proteomic technologies have made it possible to simultaneously identify multiple proteins involved in drug refractory cancers. Advances in the knowledge of dysregulation of key molecular pathways in cancer cells have enabled techniques to be developed to profile tumor cells for their genetic background, allowing selection of anticancer agents on an individual basis. The next generation of anticancer treatments might therefore be tailored according to the molecular alterations identified in tumor cells of individual patients.[53,54]

To improve the prognosis of breast cancer patients, 26 proteins that were previously determined to be important for prognosis were examined using tissue microarray. By examining 1600 cancer samples from 552 consecutive patients with early breast cancer, a set of 21 proteins was found to be the strongest independent predictor of clinical outcome in a multivariate analysis.[55] By using 2-DE and MALDI-TOF MS analysis, functional validation showed that the elevated 14-3-3σ expression contributed considerably to the observed drug resistance in breast cancer cells.[56] Its altered expression in tumors might cause clinical resistance to chemotherapy.

In nonsmall-cell lung cancer (NSCLC), patients that were not distinguishable according to tumor stages, clinical, pathologic and radiographic criteria could be separated into groups with poor or more favorable survival based on fifteen protein peaks identified using tissue MALDI of the primary tumor.[57] Patients classified as having a poor prognosis tumor had a median survival of 6 months, while patients with NSCLC in the more favorable prognosis group had a median survival of 33 months ($p < 0.0001$). On the other hand, Chen et al have reported the proteomic analysis of eIF-5A in lung adenocarcinomas using quantitative 2-DE analysis with identification by MS and 2-D immunoblots.[58] Protein expression of eIF-5A in 93 lung adenocarcinomas and 10 uninvolved lung samples were studied. Patients with higher eIF-5A protein expression showed a relatively poorer survival, suggesting the potential utility of the protein as prognostic biomarker in lung adenocarcinoma.

Using the comparative proteomic approach, several heat shock proteins (Hsps) known to complex Bcr-Abl were over-expressed in imatinib-resistant chronic myelogenous leukemia cells, showing a possible involvement of these proteins in the mechanism of drug resistance.[59] Protein HnRNPs was also found to be up-regulated in imatinib-resistant cells. These proteins have been shown to be strongly and directly related to Bcr-Abl activity.

Conclusion

Proteomics offers exciting possibilities in the realm of clinical medicine. The field foretells advances in biomarker development, improving disease diagnosis and prognostic prediction and drug development, with the ultimate promise of individualized personal therapy. Since most novel therapeutic targets are proteins, proteomic analysis potentially has a central role in patient care. Understanding the molecular basis of tumor characteristics will lead to a new era of individualized cancer therapy. Proteomic analysis therefore represents a more direct way of investigating malignancy at the individual cancer patient level. Personalized management of cancer means the prescription of specific therapeutics that best suit for an individual patient and the type of tumor. Coupled with the development of personalized therapies, the discovery of new highly sensitive and specific biomarkers for early disease detection and risk stratification holds the key to future treatment of cancer. It is becoming clear that mapping the entire networks rather than individual markers may be necessary for robust diagnostics and tailoring of therapy. Advanced proteomic platforms such as LTQ-Orbitrap MS, Fourier transform ion cyclotron resonance MS and protein arrays can generate a rapid and high resolution portrait of the proteome. Emerging novel nanotechnology strategies to amplify and harvest tumor biomarkers in vitro or in vivo will greatly enhance our ability to discover and characterize molecules for early cancer detection, subclassification and prognostic capability of current proteomic modalities. New types of proteomic technologies combined with advanced bioinformatics are currently being used to identify molecular signatures of individual tumors based on protein pathways and signaling cascades. It can be expected that analyzing the cellular circuitry of ongoing molecular networks will become a powerful clinical tool in cancer patient management. Unlike the information gathered by classical methods, high-throughput proteomic technologies can accurately inform clinicians on patient response to adjuvant therapy or those who will resist the effect of that therapy. Studies performed in cancer with high-throughput techniques have focused on tumor biology, prognosis, prediction of response to a few agents and early diagnosis. Biomarker research has become a sign of the times and the identified biomarkers may be used for clinical diagnostic or prognostic purposes. Biomarkers may also be used to help devising an optimal therapeutic treatment plan for different patient subsets and to monitor the effect of treatment. Obviously, as new biomarkers are identified and used in the clinical setting, it is hoped that these discoveries will actually translate into longer periods of disease-free survival and patient benefit directly at the bedside. In years to come, a serum or urine test for every phase of cancer may drive clinical decision making, supplementing or replacing currently existing invasive techniques.

However, issues regarding reproducibility, reliability, standardization of experimental methodologies and data analyses need to be optimized before final clinical application. Moreover, although the proteomics literature offers many examples of clinical application, few have taken the crucial step of validation of these techniques for widespread patient care. Multidisciplinary initiatives that utilize the strengths of distinct technologies, such as genomics and proteomics as well as the partnership of industry and academic research offer great potentials for improvements in patient care in the near future. Each biomarker must be critically examined and vigorously validated in different settings as we would do with any other clinical trial. Application of the most current and newest technology can result in answers to previously untestable questions but should never be a goal in itself. Although it will be a daunting task ahead, with concerted efforts we will be able to integrate the knowledge gained through genomic and proteomic analyses into clinical practice and eventually be able to provide substantial benefits to patients.

Acknowledgements

The authors wish to acknowledge the support of 2007 Chang-Jiang Scholars Program, "211" Projects, Talents Start-up Foundation of Jinan University (grant 51207040).

References

1. American Cancer Society. Cancer facts and figures. American Cancer Society 2008;
2. Srinivas PR, Kramer BS, Srivastava S. Trends in biomarker research for cancer detection. Lancet Oncol 2001; 2(11):698-704.
3. Hanash S. Disease proteomics. Nature 2003; 422:226-232.
4. Ludwig JA, Weinstein JN. Biomarkers in cancer staging, prognosis and treatment selection. Nat Rev Cancer 2005; 5:845-856.
5. Liang SL, Chan DW. Enzymes and related proteins as cancer biomarkers: A proteomic approach. Clinica Chimica Acta 2007; 381:93-97.
6. Posadas EM, Simpkins F, Liotta1 LA et al. Proteomic analysis for the early detection and rational treatment of cancer—realistic hope? Ann Oncol 2005; 16(1):16-22.
7. Latterich M, Abramovitz M, Leyland-Jones B. Proteomics: New technologies and clinical applications. Eur J Cancer 2008; 44(18):2737-2741.
8. Cho WC, Cheng CH. Oncoproteomics: current trends and future perspectives. Expert Rev Proteomics 2007; 4(3):401-410.
9. Srivastava S, Srivastava RG. Proteomics in the forefront of cancer biomarker discovery. J Proteome Res 2005; 4(4):1098-1103.
10. Steen H, Mann M. The ABC's (and XYZ's) of peptide sequencing. Nat Rev Mol Cell Biol 2004; 5(9):699-711.
11. Aebersold R, Mann M. Mass spectrometry-based proteomics. Nature 2003; 422:198-207.
12. Gorg A, Weiss W, Dunn MJ. Current two-dimensional electrophoresis technology for proteomics 2004; 4(12):3665-3685.
13. Villanueva J, Philip J, Chaparro CA et al. Correcting common errors in identifying cancer-specific serum peptide signatures. J Proteome Res 2005; 4(4):1060-1072.
14. Washburn MP, Wolters D, Yates JR. Large-scale analysis of the yeast proteome by multidimensional protein identification technology. Nat Biotechnol 2001; 19:242-247.
15. Ong SE, Blagoev B, Kratchmarova I et al. Stable isotope labelling by amino acids in cell culture, SILAC, as a simple and accurate apporach to expression proteomics. Mol Cell Proteomics 2002; 1:376-386.
16. Harsha HC, Molina H, Pandey A. Quantitative proteomics using stable isotope labeling with amino acids in cell culture. Nat Protoc 2008; 3(3):505-516.

17. Wiese S, Reidegeld KA, Meyer HE et al. Protein labeling by iTRAQ: a new tool for quantitative mass spectrometry in proteome research. Proteomics 2007; 7:340-350.
18. Choe L, D'Ascenzo M, Relkin NR et al. 8-plex quantitation of changes in cerebrospinal fluid protein expression in subjects undergoing intravenous immunoglobulin treatment for Alzheimer's disease. Proteomics 2007; 7:3651-3660.
19. Dayon L, Hainard A, Licker V et al. Relative Quantification of Proteins in Human Cerebrospinal Fluids by MS/MS Using 6-Plex Isobaric Tags. Anal Chem 2008; 80(8):2921-2931.
20. Issaq HJ, Conrads TP, Prieto DA et al. SELDI-TOF MS for diagnostic proteomics. Anal Chem 2003; 75(7):148A-155A.
21. Diamandis EP. Mass spectrometry as a diagnostic and a cancer biomarker discovery tool: opportunities and potential limitations. Mol Cell Proteomics 2004; 3(4):367-378.
22. Semmes OJ, Feng Z, Adam BL et al. Evaluation of serum protein profiling by surface-enhanced laser desorption/ionization time-of-flight mass spectrometry for the detection of prostate cancer: I. Assessment of platform reproducibility. Clin Chem 2005; 51(1):102-112.
23. Liotta LA, Espina V, Mehta AI et al. Protein microarrays: meeting analytical challenges for clinical applications. Cancer Cell 2003; 3:317-325.
24. Sheehan KM, Calvert VS, Kay EW et al. Use of reverse phase protein microarrays and reference standard development for molecular network analysis of metastatic ovarian carcinoma. Mol Cell Proteomics 2005; 4:346-355.
25. Cutler P. Protein arrays: the current state-of-the-art. Proteomics 2003; 3:3-18.
26. Ramachandran N, Hainsworth E, Bhullar B et al. Self-assembling protein microarrays. Science 2004; 305:86-90.
27. Anderson NL, Polanski M, Pieper R et al. The human plasma proteome: a nonredundant list developed by combination of four separate sources. Mol Cell Proteomics 2004; 3:311-326.
28. Sidransky D, Irizarry R, Califano JA et al. Serum protein MALDI profiling to distinguish upper aerodigestive tract cancer patients from control subjects. J Natl Cancer Inst 2003; 95:1711-1717.
29. Soltys SG, Le QT, Shi G et al. The use of plasma surface-enhanced laser desorption/ionization time-of-flight mass spectrometry proteomic patterns for detection of head and neck squamous cell cancers. Clin Cancer Res 2004; 10:4806-4812.
30. Wadsworth JT, Somers KD, Cazares LH et al. Serum protein profiles to identify head and neck cancer. Clin Cancer Res 2004; 10:1625-1632.
31. Ornstein DK, Tyson DR. Proteomics for the identification of new prostate cancer biomarkers. Urol Oncol 2006; 24:231-236.
32. Ornstein DK, Rayford W, Fusaro VA et al. Serum proteomic profiling can discriminate prostate cancer from benign prostates in men with total prostate specific antigen levels between 2.5 and 15.0 ng/ml. J Urol 2004; 172:1302-1305.
33. Wang X, Yu J, Sreekumar A et al. Autoantibody signatures in prostate cancer. N Engl J Med 2005; 353:1224-1235.
34. Bloomston M, Zhou JX, Rosemurgy AS et al. Fibrinogen gamma overexpression in pancreatic cancer identified by large-scale proteomic analysis of serum samples. Cancer Res 2006; 66:2592-2599.
35. Maciel CM, Junqueira M, Paschoal ME et al. Differential proteomic serum pattern of low molecular weight proteins expressed by adenocarcinoma lung cancer patients. J Exp Ther Oncol 2005; 5:31-38.
36. Brichory F, Beer D, Le Naour F et al. Proteomics based identification of protein gene product 9.5 as a tumor antigen that induces a humoral immune response in lung cancer. Cancer Res 2001; 61:7908-7912.
37. Xiao X, Liu D, Tang Y et al. Development of proteomic patterns for detecting lung cancer. Dis Markers 2003; 19(1):33-39.
38. Ahmed N, Oliva KT, Barker G et al. Proteomic tracking of serum protein isoforms as screening biomarkers of ovarian cancer. Proteomics 2005; 5:4625-4636.
39. Doustjalali SR, Yusof R, Govindasamy GK et al. Patients with nasopharyngeal carcinoma demonstrate enhanced serum and tissue ceruloplasmin expression. J Med Invest 2006; 53:20-28.
40. Lee IN, Chen CH, Sheu JC et al. Identification of human hepatocellular carcinoma-related biomarkers by two-dimensional difference gel electrophoresis and mass spectrometry. J Proteome Res 2005; 4:2062-2069.
41. Hermeking, H. 14-3-3 proteins and cancer biology. Semin Cancer Biol 2006; 16(3):161.
42. Lee IN, Chen CH, Sheu JC et al. Identification of complement C3a as a candidate biomarker in human chronic hepatitis C and HCV related hepatocellular carcinoma using a proteomics approach. Proteomics 2006; 6:2865-2873.
43. Melle C, Ernst G, Scheibner O et al. Identification of specific protein markers in microdissected hepatocellular carcinoma. J Proteome Res 2007; 6:306-315.

44. Xing X, Lai M, Gartner W et al. Identification of differentially expressed proteins in colorectal cancer by proteomics: down-regulation of secretagogin. Proteomics 2006; 6:2916-2923.
45. Nishimori T, Tomonaga T, Matsushita K et al. Proteomic analysis of primary esophageal squamous cell carcinoma reveals down regulation of a cell adhesion protein, periplakin. Proteomics 2006; 6:1011-1018.
46. Konner J, Schilder RJ, DeRosa FA et al. A phase II study of cetuximab/paclitaxel/carboplatin for the initial treatment of advanced-stage ovarian, primary peritoneal, or fallopian tube cancer. Gynecol Oncol 2008; 110(2):140-145.
47. Posadas EM, Liel MS, Kwitkowski V et al. A phase II and pharmacodynamic study of gefitinib in patients with refractory or recurrent epithelial ovarian cancer. Cancer 2007; 109(7):1323-1330.
48. Moen MD, McKeage K, Plosker GL et al. Imatinib: a review of its use in chronic myeloid leukaemia. Drugs 2007; 67(2): 299-320.
49. Chen Y, Shi G, Xia W et al. Identification of hypoxia-regulated proteins in head and neck cancer by proteomic and tissue array profiling. Cancer Res 2004; 64:7302-7310.
50. Rush J, Moritz A, Lee KA et al. Immunoaffinity profiling of tyrosine phosphorylation in cancer cells. Nat Biotechnol 2005; 23:94-101.
51. Souchelnytskyi S. Proteomics of TGF-beta signaling and its impact on breast cancer. Expert Rev Proteomics 2005; 2:925-935.
52. Fong L, Small EJ. Immunotherapy for prostate cancer. Curr Urol Rep 2006; 7:239-246.
53. Jain KK. Role of pharmacoproteomics in the development of personalized medicine. Pharmacogenomics 2004; 5:331-336.
54. Verrills NM, Kavallaris M. Drug resistance mechanisms in cancer cells: a proteomics perspective. Curr Opin Mol Ther 2003; 5:258-265.
55. Jacquemier J, Ginestier C, Rougemont J et al. Protein expression profiling identifies subclasses of breast cancer and predicts prognosis. Cancer Res 2005; 65:767-779.
56. Liu Y, Liu H, Han B et al. Identification of 14-3-3sigma as a contributor to drug resistance in human breast cancer cells using functional proteomic analysis. Cancer Res 2006; 66:3248-3255.
57. Yanagisawa K, Shyr Y, Xu BJ et al. Proteomic patterns of tumor subsets in nonsmall-cell lung cancer. Lancet 2003; 362:433-439.
58. Chen G, Gharib TG, Thomas DG et al. Proteomic analysis of eIF-5A in lung adenocarcinomas. Proteomics 2003; 3:496-504.
59. Ferrari G, Pastorelli R, Buchi F et al. Comparative proteomic analysis of chronic myelogenous leukemia cells: inside the mechanism of imatinib resistance. J Proteome Res 2007; 6:367-375.

Chapter 4

Metabonomics in Cancer Biomarker Discovery

Xuewu Zhang,* Gu Chen and Kaijun Xiao

Abstract

Despite the great body of knowledge about the pathogenesis of cancer and the continuous development of many drugs against cancer, the death rates for the most prevalent cancers are not significantly reduced. In order to improve treatment and reduce the mortality of cancer, a key challenge is to detect the cancer as early as possible. Over the past decade, many omics technologies (genomics, transcriptomics, proteomics and metabonomics) have been developed to identify cancer biomarkers for diagnosis, prognosis and therapy. This chapter focuses on the current applications and future challenges of metabonomics in cancer biomarker discovery. It is possible to envisage that the specialized equipment for metabonomics-based cancer diagnostics will enter the marketplace to realize personalized healthcare within the next 5-10 years.

Introduction

Currently, cancer remains a prime health concern. No robust strategy is clinically effective for primary prevention of cancer. Although we are witnessing the development of many drugs against cancer, the death rates for the most prevalent cancers are not significantly reduced. The biggest challenge in cancer control and prevention is the identification of novel biomarkers for early detection and improved therapeutic interventions to reduce mortality and morbidity rates. Unfortunately, the majority of patients are diagnosed as having cancer at late stage. Such as, 72% of lung cancer patients, 57% of colorectal cancer patients and 34% of breast cancer patients in the United States are diagnosed at late stage. If these cancers are diagnosed at early stage, the survival rate exceeds 85%.[1,2]

Biomarkers are important indicators to inform us of the physiological state of the cell at a specific time, which can prove pivotal for the identification of early cancer and people at risk of developing cancer. A plethora of omics technologies (e.g., genomics, transcriptomics, proteomics and metabonomics) are currently being used to identify molecular signatures for cancer diagnosis, prognosis and therapeutic efficacy. In this chapter, we present a comprehensive review of metabonomics in cancer biomarker discovery.

*Corresponding Author: Xuewu Zhang—College of Light Industry and Food Sciences, South China University of Technology, 381 Wushan Road, Tianhe Area, Guangzhou 510640, China. Email: snow_dance@sina.com

Omics Technologies in Cancer Biomarker Discovery, edited by Xuewu Zhang.
©2011 Landes Bioscience.

Metabonomic Technologies

Generally, there are two terms "metabonomics" and "metabolomics". Metabonomics is to investigates the fingerprint of biochemical perturbations caused by disease, drugs and toxins.[3] It is not to be confused with "metabolomics", which is defined as the comprehensive analysis of all metabolites generated in a given biological system, focusing on the measurements of metabolite concentrations and secretions in cells and tissues.[4]

Biological samples used for metabonomic analysis include biofluids (e.g., whole blood, plasma, urine, saliva, cerebrospinal fluid and prostatic fluids) and tissues (e.g., fine-needle aspirates and surgical biopsies). In general, the majority of metabonomics studies use urine or blood as samples. Urine directly reflects the metabolic status and physiological functions of the kidney or other organs. Compared with other biofluids, urine contains the highest number of water-soluble metabolites, easy to sample and requires minimal sample preparation (under normal conditions only few or no large metabolites (lipids or proteins) are present in significant concentrations). Whole blood, plasma or serum samples provide a well-established surrogate medium for metabolic physiological functions, not only for the kidney, but also for other vital organs such as liver, heart and gastrointestinal tract. Hence, blood samples can provide a "metabolic window" into the overall global metabolic response of an organism. Compared with urine, the metabolic profile of blood is less variable (rapid restoration of metabolic equilibrium) and also less dependent on, for example, dietary restrictions, but blood samples contain a complex mixture of low- and high-molecular-weight metabolites (e.g., fatty acids, glycoproteins and lipoproteins). This increases the requirements for sample extraction technology. Therefore, metabonomics studies on blood-based samples are less frequent than urine.[5]

A number of methods can produce metabolic signatures of biomaterials, including nuclear magnetic resonance (NMR), gas chromatography-mass spectrometry (GC-MS), liquid chromatography-mass spectrometry (LC-MS), capillary electrophoresis-mass spectrometry (CE-MS) and Fourier transform ion cyclotron resonance mass spectroscopy (FTICR-MS). Furthermore, a recent development in MS-based metabonomics is ultra-performance liquid chromatography (UPLC) system, with narrow columns packed with small particles (<2 μm), coupled to a hybrid quadrupole time-of-flight mass spectrometer (Q-TOF-MS).

Each technique has associated advantages and disadvantages.[6,7] For example, The advantages of NMR are: (1) nondestructive; (2) applicable to intact biomaterials, tissues and cells; (3) elucidation of molecular structure; and (4) the analytical variation has been shown to be very low for metabonomics using NMR spectroscopy, which could lead to better reproducibility of experiments than using MS.[8] On the other hand, there are some drawbacks for NMR-based metabolite profiling: (1) low resolution and sensitivity (>1 nmol metabolite for ^{1}H-NMR detection); (2) inability to detect NMR-inactive, unsuitable (e.g., O and S) or paramagnetically influenced nuclei; and (3) difficulties in absolute metabolite quantification. MS is inherently more sensitive than NMR, but it is destructive and is generally necessary to employ different separation techniques for different classes of compounds. Combination of GC with MS can be used to analyze a wide range of volatile compounds and semi-volatile compounds, its advantages include sensitivity, robustness, large linear range and large commercial and public libraries; but a big disadvantage is that derivatization is often required. In contrast, LC coupled to MS can overcome the drawback of derivatization, but limited commercial libraries are available. CE-MS has no needs for derivatization, high separation power, very small sample requirement, short analysis time and the ability to separate cations, anions and uncharged molecules in a single analytical run, but its disadvantage is poor retention time reproducibility and limited commercial libraries. For FTIR, it requires sample drying and presents extremely convoluted spectra, although no derivatization is needed. Compared with HPLC-TOF-MS system, UPLC-TOF-MS system has very short run time (2.5-3.5 min), the higher sensitivity and resolution, the signal-to-noise (S/N) ration could be increased by 5-fold.[9-14]

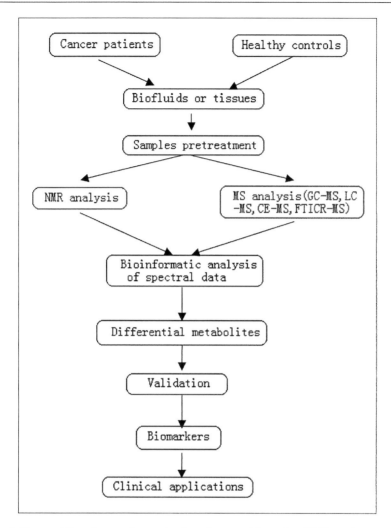

Figure 1. The workflow for metabonomics-based cancer biomarker. NMR-nuclear magnetic resonance, MS-mass spectrometry, GC-gas chromatography, LC-liquid chromatography, CE-capillary electrophoresis, FTICR-Fourier transform ion cyclotron resonance.

Current Applications and Future Challenges

The workflow for metabonomics-based cancer biomarker is shown in Figure 1. Currently, there are more and more examples of metabonomics technologies being successfully used to cancer biomarker discovery. Yang et al[15] applied HPLC-based metabonomics to diagnosis of liver cancer and found that a subset of the identified urinary nucleosides correlate better with cancer diagnosis than the traditional single tumor marker alpha-fetoprotein (AFP). Odunsi et al[16] used 1H-NMR-based metabonomics for detection of ovarian cancer and demonstrated that the sera from patients with ovarian cancer and healthy postmenopausal women could be predicted with 100% sensitivity and specificity. Al-Saffar et al[17] employed NMR technology and xenograft models to examine the effects of MN58b treatment on human colon and breast cancer cell lines. They showed that the choline kinase (CK) inhibitor MN58b reduced levels of total choline, phosphocholine (PC) and total phosphomonoesters both in vitro and in vivo, demonstrating the potential value of applying metabonomics to biomarker discovery and the identification of novel targets in a clinical translational environment. Seidel et al[18] has indicated that the simultaneous determination of modified nucleosides and creatinine in urine samples of cancer patients is a useful method for clinical diagnosis and therapy of cancer. Chen et al[19] combined desorption electrospray ionization mass spectrometry (DESI-MS) and NMR for differential metabonomics on urine samples from diseased (lung cancer) and healthy mice and successfully identified over 80 different metabolites under the condition of no sample preparation. Similarly, Beger et al[20] used NMR and HPLC-ESI-MS/MS to profile the phospholipids in plasma from pancreatic cancer patients and healthy controls and found that three phosphatidylinositols were significantly lower in pancreatic cancer patients than in healthy controls. The metabonomic model built by partial least squares discriminant function (PLS-DF) model on NMR spectra of human lipophilic plasma extracts can classify samples as pancreatic cancer group and control group with the sensitivity, specificity and overall accuracy of 98%, 94% and 96% respectively, suggesting that the changes in lipid profiles may provide a more sensitive and accurate diagnosis of pancreatic cancer than current single-biomarker methods.

It is noted that cancer is a very complex disease and involves simultaneous pathologies of multiple organ systems, in order to fulfill the dream of individualized cancer care and treatment, the combinatory use of metabonomics with other omics technologies (genomics, transcriptomics or proteomics) for cancer biomarker discovery is highly required. Currently, there has been limited work in this arena. For example, Ippolito et al[21] combined transcriptomics with metabonomics to identify features of neuroendocrine (NE) cancers associated with a poor outcome. In this study, GeneChip was firstly employed to yield a signature of 446 genes, then this signature was used for in silico metabolic reconstructions of NE cell metabolism, finally these reconstructions in turn were used to direct GC-MS/MS and LC-MS/MS analysis of metabolites in NE tumors and cell lines. The results provided a list of mRNA transcripts and metabolites indicative of a poor prognosis for various human NE cancers. Most importantly, this is a typical example showing how the hypothesis-driven approach is successfully used in metabonomic analysis, i.e., gene signatures to generate in silico prediction for metabolic pathways, then to direct MS-based metabonomic analysis for identification of metabolites. But it must be pointed out that it is not easy to establish a direct link between genes and/or proteins and metabolites: multiple mRNAs could be formed from one gene; multiple proteins from one mRNA; multiple metabolites can be formed from one enzyme; and the same metabolite can participate in many different pathways. This complicates the interpretation and integration of these omics data.

However, a major challenge in metabonomics is that the experimental metabolic profile is influenced not only by the genotype but also by age, gender, lifestyle, nutritional status, drugs, stress, physical activity, etc. To minimize the variations in studies with humans, the standardization of the entire experimental process is highly required, such as automation of all sample preparation steps, using standardized diet, avoiding any drug, stopping any vigorous activity, excluding smokers and so on. A recent report reveals a "natural", stable over time and invariant metabolic profile for each person, although the existence of human metabolic variations resulting from various dietary patterns.[22] This provides the possibility of eliminating the day-to-day "noise" of the individual metabolic fingerprint and opens new perspectives to metabonomic studies for biomarker discovery.

The second challenge in metabonomics is gut microbiota-host metabolic interactions, such a interaction between the microbiome and humans makes the human become a "superorganism", there are more than 400 microbial species in the large-bowel microflora of healthy humans.[3] Recently, based on the plasma and urine metabolic NMR profile of mouse, Dumas et al[23] investigated the metabolic relationship between gut microflora and host cometabolic phenotypes. They found that the urinary excretion of methylamines from the precursor choline was directly related to microflora metabolism, demonstrating significant interaction between the mammalian host and microbiota metabolism. As such, the significant metabolic signals produced by gut microflora may "swamp" the true metabonomic signals elicited by cancer, leading to the altered metabonome in human biofluids and possible misidentification of biomarkers.

Finally, a major challenge in metabonomics is the assignment of metabolites. In order to rapidly identify metabolites, powerful bioinformatic tools, databases and data analysis algorithm are highly required. For database, there is limited metabonome database available, especially in the field of cancer research. For example, there are two MS-based metabonome databases, one is GMD, a comprehensive GC-MS metabolite profiles libraries from mammals, corynebacteriae and major plant species;[24,25] another is METLIN, which includes an annotated list of known metabolite structural information that is easily cross-correlated with its catalogue of high-resolution Fourier transform mass spectrometry (FTMS) spectra, tandem mass spectrometry (MS/MS) spectra and LC/MS data.[26] For metabonomic data analysis, a wide range of statistical and machine-learning algorithms have been developed. For example, Ebbels et al[27] presented a CLOUDS method for analyzing NMR spectra of urinary metabonome. Cloarec et al[28] described an exploratory approach, statistical total correlation spectroscopy analysis method, for biomarker identification from metabolic NMR data sets. Smilde et al[29] used ANOVA-simultaneous component analysis for examining the effect of vitamin C on the development of osteoarthritis. Broeckling et al[30] developed a data extraction tool for MS-based metabonomics. Recently, Dieterle et al[31] developed a metabolite projection analysis method, which allows easy and fast tentative assignment of all structures of metabolites significantly changed in the urinary metabonomics. Although such a method is based on NMR spectra at present, it is conceivable to extend the method to MS-based metabonomics when the corresponding spectral databases are available.

Conclusion

Metabonomics are showing the potential for cancer diagnosis and possibly even prognosis. The complementary analytical methodologies, such as NMR- and MS-based metabolic analysis, will significantly contribute to the identification of metabolic biomarkers. In order to reduce the biased results, the standardization of experimental protocols and automation of metabonomics facility are highly required. Much attention should also be paid to the validation of defined metabolites and cross-validation of key metabolites obtained by NMR or MS with other quantitative assays. In particular, the combination of metabonomics datasets with other omics such as genomics or proteomics datasets will certainly provide a more complete view into the molecular pathways of systems biology. It is easy to imagine that within the next 5-10 years the world will see a revolution in healthcare. The personalized medicine and healthcare will become reality and the specialized diagnostics procedures based on metabolic profile or its combinational profiles with genomics and proteomics will be on the market.

References

1. Gloeckler-Ries LA, Reichman ME, Lewis DR et al. Cancer Survival and Incidence from the Surveillance, Epidemiology and End Results (SEER) Program. Oncologist 2003; 8(6):541-552.
2. Semmes OJ, Feng Z, Adam BL et al. Evaluation of serum protein profiling by surface-enhanced laser desorption/ionization time-of-flight mass spectrometry for the detection of prostate cancer: I. Assessment of platform reproducibility. Clin Chem 2005; 51(1):102-112.
3. Goodacre R. Metabolomics of a superorganism. Journal of Nutrition 2007; 137(1):259S-266S Suppl.
4. Fiehn O. Metabolomics—the link between genotypes and phenotypes. Plant Mol Biol 2002; 48(1-2):155-171.
5. Serkova NJ, Niemann CU Pattern recognition and biomarker validation using quantitative 1H-NMR-based metabolomics. Expert Rev Mol Diagn 2006; 6(5):717-731.
6. Shulaev V. Mtabolomics technology and bioinformatics. Brief Bioinform 2006; 7(2):128-139.
7. Zhang XW, Wei D, Yap YL et al. Mass spectrometry-based "omics" technologies in cancer diagnostics. Mass Spectrom Rev 2007; 26:403-431.
8. Lindon JC, Holmes E, Nicholson JK. Metabonomics and its role in drug development and disease diagnosis. Expert Rev Mol Diagn 2004; 4(2):189-199.
9. Castro-Perez J, Plumb R, Granger JH et al. Increasing throughput and information content for in vitro drug metabolism experiments using ultra-performance liquid chromatography coupled to a quadrupole time-of-flight mass spectrometer. Rapid Commun Mass Spectrom 2005; 19(6):843-848.

10. Crockford DJ, Lindon JC, Cloarec O et al. Statistical Search Space Reduction and Two-Dimensional Data Display Approaches for UPLC-MS in Biomarker Discovery and Pathway Analysis. Anal Chem 2006; 78(13):4398-4408.

11. de Villiers A, Lestremau F, Szucs R et al. Evaluation of ultra performance liquid chromatography Part I. Possibilities and limitations. J Chromatogr A 2006; 1127(1-2):60-69.

12. Plumb RS, Granger JH, Stumpf CL et al. A rapid screening approach to metabonomics using UPLC and oa-TOF mass spectrometry: application to age, gender and diurnal variation in normal/Zucker obese rats and black, white and nude mice. Analyst. 2005; 130(6):844-849.

13. Plumb RS, Johnson KA, Rainville P et al. UPLC/MS(E); a new approach for generating molecular fragment information for biomarker structure elucidation. Rapid Commun Mass Spectrom 2006; 20(14):2234.

14. O'Connor D, Mortishire-Smith R. High-throughput bioanalysis with simultaneous acquisition of metabolic route data using ultra performance liquid chromatography coupled with time-of-flight mass spectrometry. Anal Bioanal Chem 2006; 385(1):114-121.

15. Yang J, Xu G, Zheng Y et al. Diagnosis of liver cancer using HPLC-based metabonomics avoiding false-positive result from hepatitis and hepatocirrhosis diseases. J Chromatogr B Analyt Technol Biomed Life Sci 2004; 813(1-2):59-65.

16. Odunsi K, Wollman RM, Ambrosone CB et al. Detection of epithelial ovarian cancer using 1H-NMR-based metabonomics. Int J Cancer 2005; 113(5):782-788.

17. Al-Saffar NMS, Troy H, de Molina AR et al. Noninvasive magnetic resonance spectroscopic pharmacodynamic markers of the choline kinase inhibitor MN58b in human carcinoma models. Cancer Res 2006; 66(1):427-434.

18. Seidel A et al. Modified nucleosides: an accurate tumor marker for clinical diagnosis of cancer, early detection and therapy control. Br J Cancer 2006; 94(11):1726-1733.

19. Chen H, Pan Z, Talaty N et al. Combining desorption electrospray ionization mass spectrometry and nuclear magnetic resonance for differential metabolomics without sample preparation. Rapid Commun Mass Spectrom 2006; 20(10):1577-1584.

20. Beger RD, Schnackenberg LK, Holland RD et al. Metabonomic models of human pancreatic cancer using 1D proton NMR spectra of lipids in plasma. Metabolomics 2006; 2(3):125-134.

21. Ippolito JE, Xu J, Jain S et al. An integrated functional genomics and metabolomics approach for defining poor prognosis in human neuroendocrine cancers. Proc Natl Acad Sci USA 2005; 102(28):9901-9906.

22. Assfalg M, Bertini I, Colangiuli D et al. Evidence of different metabolic phenotypes in humans. Proc Natl Acad Sci USA 2008; 105(5):1420-1424.

23. Dumas ME, Barton RH, Toye A et al. Metabolic profiling reveals a contribution of gut microbiota to fatty liver phenotype in insulin resistant mice. Proc Natl Acad Sci USA 2006; 103(33):12511-12516.

24. Kopka J, Schauer N, Krueger S et al. GMD@CSB.DB: the Golm Metabolome Database. Bioinform 2005; 21(8):1635-1638.

25. Schauer N, Steinhauser D, Strelkov S et al. GC-MS libraries for the rapid identification of metabolites in complex biological samples. FEBS Lett 2005; 579:1332-1337.

26. Smith CA, O'Maille G, Want EJ et al. METLIN—A metabolite mass spectral database. Ther Drug Monit 2006; 27(6):747-751.

27. Ebbels T, Keun H, Beckonert O et al. Toxicity classification from metabonomic data using a density superposition approach: 'CLOUDS'. Analytica Chimica Acta 2003; 490(1-2):109-122.

28. Cloarec O, Dumas ME, Craig A et al. Statistical total correlation spectroscopy: An exploratory approach for latent biomarker identification from metabolic H-1 NMR data sets. Anal Chem 2005; 77(5):1282-1289.

29. Smilde AK, Jansen JJ, Hoefsloot HCJ et al. ANOVA-simultaneous component analysis (ASCA): a new tool for analyzing designed metabolomics data. Bioinform 2005; 21(13):3043-3048.

30. Broeckling CD, Reddy IR, Duran AL et al. MET-IDEA: Data extraction tool for mass spectrometry-based metabolomics. Anal Chem 2006; 78(13):4334-4341.

31. Dieterle F, Ross A, Schlotterbeck G et al. Metabolite projection analysis for fast identification of metabolites in metabonomics. Application in an amiodarone study. Anal Chem 2006; 78(11):3551-3561.

CHAPTER 5

Peptidomics in Cancer Biomarker Discovery

Lei Shi*

Abstract

The emergence of novel technologies allows researchers to comprehensively analyze genomes, transcriptomes and proteomes in health and disease. Peptidomics is the field that deals with the comprehensive qualitative and quantitative analysis of peptides in biological systems. The term peptidomics is relatively new and was first mentioned in the literature less than eight years ago. The science of peptidomics is developing rapidly. Analysis of peptides in biological samples promises to provide diagnostic and prognostic information for cancer. The information that is expected from peptidomics may soon exert a dramatic change in the pace of cancer research. The extraordinary power of mass spectrometry in identifying and quantifying peptides in complex biological mixtures offers opportunities for developing novel technologies for diagnosis of cancer.

Introduction

The finalization of the human genome projects has entered biology into a new era. Although the sequences of some genes are known at present, for most genes very little is known about their function and their interaction. This has led to the introduction of "postgenomic" techniques to discover the proteins that are regulated in certain physiological processes.

"Proteomics" attempts to identify all the proteins and their posttranslational modifications in a certain organism or tissue. The most common approach in proteomic studies is to separate and visualize all the proteins by two-dimensional (2D) gel electrophoresis and to subsequently identify the expressed proteins by mass spectrometric techniques.

However, there are several limitations to the use of two-dimensional polyacrylamide gel electrophoresis (2D-PAGE) as a proteome visualization technique. One of the major constraints is the limitation of the size of proteins. Proteins of a molecular mass lower than 10 kDa can not be studied. Nevertheless, this mass region contains the very important groups of peptides.

Peptides play a central role in many biological processes and some classes of such biologically active peptides, e.g., hormones, cytokines and growth factors, but only a very small number of scientific manuscripts deal with an integrated analysis of the peptide content of an organism, body fluid, tissue or cell. This type of analysis was named "peptidomics". The term peptidomics is relatively new and was first mentioned in the literature less than eight years ago.

Cancer remains a major public health challenge despite progress in detection and therapy. A large portion of the United States population will develop cancer during their lifetime,[1] with 500000 individuals dying annually from the disease.[2] The race to obtain control over the disease process is gaining speed and focus. From biotechnology to chemistry, from applied physics to software, increasing resources are being brought to bear on the goals of prevention and reducing mortality.

*Lei Shi—College of Light Industry and Food Sciences, South China University of Technology, 381 Wushan Road, Guangzhou 510640, China. Email: leishi@scut.edu.cn

Omics Technologies in Cancer Biomarker Discovery, edited by Xuewu Zhang.
©2011 Landes Bioscience.

Innovations and applications of biotechnology have allowed the exploitation of biological processes in an effort to study pathogenesis at the molecular level.

Novel technologies that are designed to advance the molecular analyses of healthy and diseased human cells are poised to revolutionize the field of health and disease. Advances in the fields of Peptidomics are hoped to provide insights into the molecular complexity of the disease process and thus enable the development of tools to help in treatment as well as in detection and prevention. Among the important tools critical to detection, diagnosis, treatment, monitoring and prognosis are biomarkers.

Biomarkers are biological molecules that are indicators of physiologic state and also of change during a disease process.[3] The utility of a biomarker lies in its ability to provide an early indication of the disease, to monitor disease progression, to provide ease of detection and to provide a factor measurable across populations.

The human genome[4,5] has set the pace for biomarker discovery and provided the impetus for the next level of molecular inquiry. Peptidomics deals with the comprehensive qualitative and quantitative analysis of peptides in biological samples.[6] Logically and to the uninitiated, peptidomics is the specification of the complement of peptides of a cell, organelle, tissue or organism. These peptides are either intact small molecules, such as hormones, cytokines and growth factors, peptides that are released form larger protein precursors during protein processing or they may represent degradation products of proteolytic activity. Thus, in biological fluids, peptides represent protein synthesis, processing and degradation. Since the amount and repertoire of peptides in the circulation change dynamically according to the physiological or pathological state of an individual, it is possible that comprehensive peptide analysis may lead to discovery of novel biomarkers or to new diagnostic approaches.[7]

Peptidomics Technology

A proteomic analysis of a sample usually consists of four steps. These are: extraction of the proteins, their separation, detection and finally identification of the individual, separated proteins. The separation and quantitative detection of proteins is routinely performed by 2D gel electrophoresis and staining of the gel. Protein identification is performed by the mass spectrometric analysis of a tryptic digest of the individual proteins through peptide mass fingerprinting or tandem mass spectrometry.

It has been demonstrated that peptides can be analyzed by mass spectrometry which requiring very little or no manipulations of the samples, just by placing the tissues directly on the matrix assisted laser description/ionization (MALDI) target plate and applying the matrix solution[8] or, in the case of electrospray mass spectrometry (ESI), by a brief extraction in the spraying solvent.[8,9] However, in many case, the complexity of the peptide sample requires a separation prior to the analysis of the sample. Several mass spectrometric techniques have been used to detect and identify neuropeptides from tissue extracts using a very limited amount of starting material. Both approaches with matrix assisted laser desorption time of flight (MALDI-TOF) and electrospray mass spectrometry have proven to be successful.[8,10]

Powerful analytical technologies allowed identification of numerous and previously unknown peptides in the circulation. The extraordinary power of mass spectrometry in identifying and quantifying peptides in complex biological mixtures offers opportunities for developing novel technologies for diagnosis of cancer and other diseases.

It should be pointed out that a pivotal technological challenge for peptidome analysis is how to efficiently extract the peptides from highly complex samples like human serum. To reduce the suppression of endogenous peptides by high abundance proteins, several methodologies are currently used to enrich peptides from biological samples. For example, centrifugal ultrafiltration, solid-phase extraction (SPE), acetonitrile (ACN) precipitation, size-exclusion chromatography, magnetic beads with defined surface functionalities (hydrophobic interaction, cation exchange and metal ion affinity), etc.

A "New Era" in Cancer Diagnostics

In one of the first attempts to use serum proteomic (including peptidomic) profiling, in combination with surface enhanced laser desorbtion-ionization time-of-flight mass spectrometry, Petricoin et al reported outstanding sensitivities and specificities for early ovarian cancer diagnosis by using peaks of unknown identity.[11]

Subsequently, others developed similar technologies for diagnosing numerous other cancer types, with claimed sensitivities and specificities that far exceed those achieved with the classical cancer biomarkers.[12] It was thus postulated that a "new era" in cancer diagnostics had emerged, in which serum proteomic/peptidomic profiles would fulfill the goal of early cancer detection. Unfortunately, other groups soon identified major methodological and bioinformatic artifacts and biases.[13-16]

It is clear that sample collection, storage and processing procedures can produce proteomic patterns that could overshadow those generated by the presence of disease.[17,18] Naturally, the enthusiasm of using these technologies for diagnostics has declined, in part due to the inability of other groups to reproduce or validate the originally published data.[19,20]

In 2003, Marshall et al claimed that peptides from the sera of normal individuals and patients who suffered myocardial infarction (MI) can produce MALDI-TOF patterns that provide an accurate diagnosis of MI.[21] It is shown that the spectral patterns mainly originated from the cleavage of complement C3 alpha chain to release the C3f peptide and from cleavage of fibrinogen to release peptide A.

The fibrinogen peptide A and complement C3f peptides were in turn progressively truncated by amino peptidases to produce two families of fragments that formed the characteristic spectral pattern of MI. They have shown that the peptide patterns in serum reflected the balance of disease-specific protease and aminopeptidase activity ex-vivo. Around the same time, Liotta et al postulated that serum and/or plasma contains a large repertoire of different peptides which are bound to high abundance proteins such as albumin and are thus protected from clearance by the kidneys.[26]

They further hypothesized that these peptides may have important diagnostic value. Apparently, there are various classes of peptides in serum, some of them of relatively high abundance. The low molecular weight peptides may carry some important diagnostic information. The diagnostic potential of low molecular weight peptides has recently been explored by Lopez et al for diagnosis of Alzheimer's disease[23] and by Lowenthal et al for diagnosis of ovarian carcinoma.[24]

Koomen et al provided important information on the generation of the serum peptidome by using peptide extraction, fractionation and characterization by liquid chromatography coupled to MALDI tandem mass spectrometry.[25] These authors can identify more than 250 peptides in plasma and demonstrated that they originated from a surprisingly small number of proteins. The proteins were very common and of high abundance, including fibrinogen, complement components, antiproteases, apoliproproteins, acute phase reactants and carrier proteins. It is postulated that initial endoproteolytic cleavages of these abundant proteins occur by common enzymes such as thrombin, plasmin and complement proteins, followed by aminopeptidase and carboxypeptidase exoprotease processing.

Recently, Villanueva et al suggested a novel way of diagnosing cancer by using peptidomic analysis, combined with MALDI-TOF mass spectrometry.[26] The method received enthusiastic endorsement by some.[27] The procedure of Villanueva and identified by surface-enhanced laser desorption/ionization (SELDI-TOF) mass spectrometry, do not appear upon reevaluation by confirmatory techniques such as enzyme-linked immunosorbent assay (ELISA) to have much value in diagnostics.

As postulated earlier,[17-15] putative new cancer biomarkers, including peptides, are likely to be present in biological fluids at extremely low concentrations, necessitating identification by quantitative, highly sensitive and reproducible techniques. Convincing data for identifying low abundance, tumor-derived peptides for diagnostic purposes are currently lacking.

The paper by Villanueva et al suffers from important design biases which put their findings in question.[22] For example, they selected young men and women as the control group in contrast to the much older patient groups. It is thus entirely possible that their findings are related more to the age of the patients than to the presence of cancer. Despite the caution by Ransohoff in designing studies of this kind,[11] high-profile papers like this one are still compromised by avoidable bias. Villanueva et al postulated that tumor-related exoprotease activity may be responsible for the observed findings. However, this hypothesis is inconsistent with their data. For example, a series of peptides was informative for prostate, bladder and breast cancer. These peptides originate from one parental peptide that is further processed by a putative aminopeptidase. If this aminopeptidase was increased due to tumor presence, then one would expect that the whole series of daughter peptides would increase in all types of cancer. However, while a number of peptides increase and others do not change, in prostate and bladder cancer, three of these peptides actually decreased in breast cancer. This puts into question the hypothesis that tumor-derived amino peptidases are likely generating the diagnostic information.

It is now time to approach proteomic and peptidomic profiling for cancer diagnosis with increased scrutiny, to avoid biases and move to the next step.

Conclusion

The methods are affected by sample collection and storage conditions, some parameters are thoroughly examined. Studies should be designed carefully so that biological and bioinformatic biases are avoided. The effects of variables such as age, gender, common drug ingestion, dietary habits, exercise, etc. should be studied, to establish if the methods are robust enough for routine testing. Since some of the methods are dependent on ex-vivo proteolysis and on pathways such as complement and coagulation, the effects of coagulopathies and inflammatory conditions should be reported. When deriving clinical sensitivities and specificities, patients with early stage, in addition to late stage disease and age-matched controls with nonmalignant inflammatory or other related conditions should be included in the study. The original hypothesis should not be conflicting with the data provided. In cases where putative proteolytic activity is implicated as the discriminatory factor, it is important to characterize the proteolytic activity, to strengthen the hypothesis. The identified biomarkers or panels should be examined with other standard parameters, i.e., correlation with tumor burden and other clinic pathological variables, change of biomarker concentration after treatment or postsurgery, etc. All cancer biomarkers used at the clinic today are increased in the circulation since they are derived from tumor cells. Biomarker concentration decrease with tumor burden increase should be interpreted with caution and explained biologically, where possible.

These new diagnostic methods will be accepted at the clinic only when it is shown in large and well-designed validation studies that they have value for patient care. In such studies, the originally derived algorithms and multiparametric schemes proposed by the authors should be applied to patients who are assessed blindly and with samples originating from different institutions under standardized sample collection and storage conditions. It will be highly important that validation studies with negative results are published, so that the scientific literature is cleaned from data that do not represent real advances in the field.

References

1. Chaurand P, DaGue BB, Pearsall RS et al. Profiling proteins from azoxymethane-induced colon tumors at the molecular level by matrix-assisted laser desorption/ionization mass spectrometry. Proteomics 2001; 1320-1326.
2. American Cancer Society.Cancer facts and figures. Atlanta: American Cancer Society, 1996.
3. Srinivas PR, Kramer BS, Srivastava S. Trends in biomarker research for cancer detection. Lancet Oncol 2001; 2:698-704.
4. Lander ES, Linton LM, Birren B et al. Initial sequencing and analysis of the human genome. Nature 2001; 409:860-921.
5. Venter JC, Adams MD, Myers EW et al. The sequence of the human genome. Science 2001; 291:1304-1351.

6. Schulte I, Tammen H, Selle H et al. Peptides in body fluids and tissues as markers of disease. Expert Rev Mol Diagn 2005; 5:145-157.

7. Schrader M, Schulz-Knappe P. Peptidomics technologies for human body fluids. Trends Biotechnol 2001; 19:S55-60.

8. Clynen E, Baggerman G, Veelaert D et al. Peptidomics of the pars intercerebralis-corpus cardiacum complex of the migratory locust, Locusta migratoria. J Biochem 2001; 268:1929.

9. Kollisch GV, Lorenz MW, Kellner R et al. Structure elucidation and biological activity of an unusual adipokinetic hormone from corpora cardiaca of the butterfly, Vanessa cardui. Eur J Biochem 2000; 267:5502.

10. Predel R, Kellner R, Baggerman R et al. Identification of novel periviscerokinins from single neurohaemal release sites in insects. Eur J Biochem 2000; 267:3869.

11. Petricoin EF, III, Ardekani AM, Hitt BA et al. Use of proteomic patterns in serum to identify ovarian cancer. Lancet 2002; 359:572-577.

12. Wulfkuhle JD, Liotta LA, Petricoin EF. Proteomic applications for the early detection of cancer. Nat. Rev Cancer 2003; 3:267-275.

13. Diamandis EP. Proteomic patterns in biological fluids: Do they represent the future of cancer diagnostics? Clin Chem 2003; 49:1272-1278.

14. Diamandis EP. Analysis of serum proteomic patterns for early cancer diagnosis: Drawing attention to potential problems. J Natl Cancer Inst 2004; 96:353-356.

15. Diamandis EP, Merwe DE. Plasma protein profiling by mass spectrometry for cancer diagnosis: opportunities and limitations. Clin Cancer Res 2005; 11:963-965.

16. Baggerly KA, Morris JS, Edmonson SR et al. Signal innoise evaluating reported reproducibility of serumproteomic tests for ovarian cancer. J Natl Cancer Inst 2005; 97:307-309.

17. Banks RE, Stanley A, Cairns DA et al. Influences of blood sample processing on low-molecular-weight proteome identified by surface-enhanced laser desorption/ionization mass spectrometry. Clin Chem 2005; 51:1637-1649.

18. Karsan A, Eigl BJ, Flibotte et al. Analytical and preanalytical biases in serum proteomic pattern analysis for breast cancer diagnosis. Clin Chem 2005; 51:1525-1528.

19. Ransohoff DF. Bias as a threat to the validity of cancer molecular-marker research. Nat Rev Cancer 2005; 5:152-149.

20. Ransohoff DF. Lessons from controversy: ovarian cancer screening and serum proteomics. J Natl Cancer Inst 2005; 97:315-319.

21. Marshall J, Kupchak P, Zhu W et al. Processing of serum proteins underlies the mass spectral fingerprinting of myocardial infarction. J Proteome Res 2003, 2:361-372.

22. Liotta LA, Ferrari M, Petricoinm E. Clinical proteomics: written in blood. Nature 2003; 425:905.

23. Lopez MF, Mikulskis A, Kuzdzal S et al. High-resolution serum proteomic profiling of Alzheimer disease samples reveals disease-specific, carrierprotein-bound mass signatures. Clin Chem 2005; 51:1946-1954.

24. Lowenthal MS, Mehta AI, Frogale K et al. Analysis of albumin-associated peptides and proteins from ovarian cancer patients. Clin Chem 2005; 51:1933-1945.

25. Koomen JM, Li D, Xiao L et al. Direct tandem mass spectrometry reveals limitations in protein profiling experiments for plasma biomarker discovery. J Proteome Res 2005; 4:972-981.

26. Villanueva J, Shaffer DR, Philip J et al. Differential exoprotease activities confer tumor-specific serum peptidome patterns. J Clin Invest 2006; 116:271-284.

27. Liotta LA, Petricoin EF. Serum peptidome for cancer detection: spinning biologic trash into diagnostic gold. J Clin Invest 2006; 116:25-30.

CHAPTER 6

Phosphoproteomics in Cancer Biomarker Discovery

Li-Rong Yu,* Yuan Gao and Donna L. Mendrick

Abstract

B iomarkers are needed for personalized cancer therapy including early detection of cancer, disease stratification and monitoring therapeutic outcomes. Recent advances in phosphoprotein enrichment and mass spectrometry (MS) technologies for quantitative phosphoproteome analysis in both global and targeted approaches provide great opportunities in the identification and validation of phosphoprotein biomarkers. These technologies are useful tools for the identification of phosphosignatures unique to a specific type or subtype of cancer and drug responsive biomarkers for efficacy and toxicity. This chapter describes some examples of phosphoproteins serving as cancer biomarkers, the advances of phosphoprotein/phosphopeptide enrichment, MS technologies for global and targeted quantitative phosphoproteome analysis, biomarker validation and qualification and some interesting applications.

Introduction

The advancements in cancer biology such as understanding the mechanisms underlying tumorigenesis have significantly changed the concept of cancer therapeutics. As a result of searching for crucial cancer-causing oncogenes and tumor-suppressor genes in the 1980s,[1,2] the concept of targeting a specific biological macromolecule (e.g., kinase) or molecular event for therapeutics was developed. Such molecule-targeted therapeutics could provide better outcomes than standard radiation and chemotherapy that have severe side effects and a limited capacity to discriminate between healthy and cancerous cells.[3] Clinical trials have demonstrated efficacy of some drugs that target specific kinases that play critical roles in oncogenic pathways, which has led to the approval of some kinase inhibitors by the U.S. Food and Drug Administration. Accordingly, the focus of drug discovery and development in the past two decades has been shifted from nonspecific chemotherapeutics to rationally designed drugs that target cancer-specific molecules and pathways.[3]

Although targeted cancer treatment has demonstrated great promise of eliminating this type of disease, often the cancer eventually relapses. One of the reasons is that most patients are already at an advanced stage of disease when they are hospitalized and diagnosed.[4] To benefit patients and reduce mortality, early cancer detection is critical. For example, the five-year survival rate is approximately 20% for breast cancer patients diagnosed with distant disease while the rate is 90% or more for those detected at early stages when the tumor is localized.[4] Furthermore, the chance of survival for 10 years is greater than 80% for the breast cancer patients treated at an early stage.[4] However, early diagnosis relies on clinically validated biomarkers with high specificity and sensitivity. While lots of research efforts have been made in recent years to discover disease

*Corresponding Author: Li-Rong Yu—Center for Proteomics, Division of Systems Biology, National Center for Toxicological Research, U.S. Food and Drug Administration, Jefferson, Arkansas, USA. Email: lirong.yu@fda.hhs.gov

Omics Technologies in Cancer Biomarker Discovery, edited by Xuewu Zhang.
©2011 Landes Bioscience.

markers,[5,6] the path from discovery to clinical use is still in its early stage[7] and much more work is needed before putative markers are proven to be useful in clinical practice. It is believed that cancer biomarkers can be causal or consequential in nature and can be macromolecules or metabolites. Specific molecules involved in oncogenic pathways, such as phosphoproteins, would be of particular interest since phosphorylation states of proteins could serve as useful signatures to facilitate therapeutic decision-making and prediction of clinical outcomes of a specific targeted therapy.

Various proteomics technologies have been developed recently[8,9] to profile proteins from tumor cell lines, tissues, sera and plasma for the identification of cancer biomarkers.[5,6] These technologies include mass spectrometry (MS)-based and nonMS-based tools as exemplified by Multidimensional Protein Identification Technology (MudPIT)[10] and protein microarrays,[11] respectively. The term "Oncoproteomics" has been introduced to describe the study of proteins and their functions in cancer cells by proteomic technologies.[12] A majority of biomarker research has focused on the measurement of global protein abundance changes in sera or plasma. Nonetheless, studies have begun to search for protein modification changes (e.g., phosphorylation and glycosylation) relevant to disease development.[13,14] While phosphorylation analysis of individual proteins has been conducted broadly in cell biology research, it is still relatively young for the analysis of protein phosphorylation at the proteome level (i.e., phosphoproteomics) in the field of clinical proteomics. Recent advancements in phosphoproteomic technologies[15] provide great promise and opportunities in applying such tools to the identification of phosphoproteins as cancer biomarkers. Although there are many approaches to the study of phosphoproteins, here we mainly focus on MS-based phosphoproteome analysis for the development of phosphoprotein cancer biomarkers. Challenges in this field are discussed below and include (a) sample preparation and processing, (b) advantages and disadvantages of available approaches, (c) discovery of biomarkers and (d) clinical qualification of selected markers.

Phosphoproteins as Cancer Biomarkers

It is well-understood and broadly accepted that cancer is a multi-factorial disease, which is generally considered to be genetic.[16] However it can be functionally classified as a proteomic disease.[17] While modifications at the genetic (e.g., mutation and amplification) and epigenetic levels (e.g., promoter hypermethylation) are considered to contribute to cancer initiation and development,[16,18] these changes are closely associated or co-exist with functional changes (e.g., loss of function) of proteins. Such changes ultimately lead to deregulation of signal transduction pathways in which proteins and DNA (i.e., genes) interact actively and diversely to promote tumorigenesis.[19] Both genomic and proteomic changes induce further genome instability, which in turn modifies genes and proteins to confer cancer progression. The most well-known mechanism in the regulation of oncogenic pathways is protein phosphorylation and de-phosphorylation via protein kinases and phosphatases, respectively. In addition, it has been widely demonstrated that phosphorylation is a major type of modification to regulate a variety of functions for structural proteins, co-activators, cosuppressors, transcription factors, etc. Identification of phosphoprotein signatures specific to a certain type of cancer would facilitate disease diagnosis and classify patients that require personalized pathway-directed cancer therapy. Such signatures could be comprised of single or multiple phosphosites and/or phosphoproteins. A panel of biomarkers likely will be predominant over single biomarkers and could exist within the same or different signaling pathways (Fig. 1).

Great efforts have been made to identify cancer biomarkers using proteomic technologies[5,6] and more than one thousand proteins have been identified as being differentially expressed in human cancer.[20] Many of these biomarkers are phosphoproteins, especially kinases. Examples of regulatory-approved phosphoproteins as cancer biomarkers include epidermal growth factor receptor (EGFR) for colon cancer, Kit for gastrointestinal stromal tumors (GIST) and HER2 (human EGFR2) for breast cancer.[21] All of these are receptor tyrosine kinases (RTKs) and the sites and degree of tyrosine phosphorylation might be distinct in different states of disease and between subjects.[22] Kit has been used for diagnosis and therapy selection, while the other two kinases have

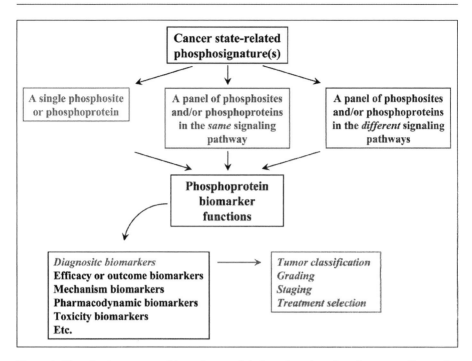

Figure 1. Phosphosignatures as biomarkers and their various functions in cancer diagnostics and therapeutics.

been clinically used not only for therapy selection but also for prognosis, monitoring therapeutic effects and disease recurrence.[21] Approximately 30% of patients with invasive breast cancer have *HER2/neu* amplification,[23] which is correlated with reduced survival rate.[24] It is expected that many more phosphoproteins, including kinases, can be discovered and serve as biomarkers for cancer diagnosis and prognosis.

While the utility of phosphoprotein biomarkers in cancer detection and diagnosis is evident, it should be noted that biomarkers can be used in the whole range of patient care and drug development (Fig. 1). Different types of biomarkers, including those related to disease, surrogate endpoints, efficacy or outcome, mechanism, pharmacodynamics, drug target, toxicity and bridging or translational biomarkers[25] could be identified from phosphorylated protein species. To specify the use of biomarkers at different stages of cancer diagnosis, Ludwig and Weinstein[21] classified biomarkers in detail for tumor classification, grading, staging, prognosis and treatment selection. They also noted that markers might be used for risk assessment and screening prior to diagnosis. Three examples are provided here. Taguchi et al defined a MALDI-MS method to assess the outcome of EGFR inhibitors gefitinib (Iressa™, AstraZeneca) and erlotinib (Tarceva®, Genentech) for the treatment of nonsmall-cell lung cancer (NSCLC). The eight distinct *m/z* features they found could classify NSCLC patients that responded well or poorly upon treatment with EGFR inhibitors.[26] In an another example, the mammalian target of rapamycin (mTOR) is a serine/threonine (S/T) kinase that functions as a master switch between catabolic and anabolic metabolism and has been the target for anticancer drug development. Preclinical studies have revealed that mTOR inhibition has a correlation with inactivation of ribosomal protein S6 kinase 1 (S6K1) and the phosphorylation status of S6K1 can be used as a biomarker for mTOR inhibitors.[27] Sorafenib (Nexavar®, Bayer HealthCare Pharmaceuticals) is a multikinase inhibitor that affects tumor signaling and tumor vasculature. A recent Phase II trial of this drug indicated that high baseline pERK levels could be indicative of the hepatocellular carcinoma response to sorafenib.[3] These findings suggest that phosphoproteins or the degree of phosphorylation of a particular protein can serve as biomarkers and play important roles in the assessment of drug response to targeted cancer therapies. Discovery of such phosphoprotein biomarkers would provide invaluable tools for cancer drug development and therapeutic monitoring.

Sample Preparation for Phosphoprotein/Phosphopeptide Enrichment

Protein phosphorylation analysis can be performed currently at a large scale and in a somewhat high-throughput fashion taking advantage of great improvements in recent years in MS and phosphopeptide enrichment. Nonetheless, there are still many challenges in the detection and identification of phosphorylated proteins and their corresponding phosphorylated sites mainly owing to the low stoichiometric nature of protein phosphorylation in cells and poor fragmentation of certain classes of phosphopeptides during mass spectrometry analysis.[28,29] The phosphoproteins/phosphopeptides must be enriched prior to downstream MS measurement for phosphoproteome characterization and quantitative analysis. Some approaches have been or are being developed for the enrichment of phosphorylated peptides and proteins.

The technologies for the enrichment of phosphorylated proteins or peptides can generally be classified into four basic strategies: chemical modification of phosphorylated residues, immunoaffinity separation, ion exchange chromatography and metal ion affinity chromatography (Table 1). In the methods using chemical modifications, β-elimination followed by Michael addition is usually employed to replace the O-phosphate moieties of the phosphoserine- (pSer) and phosphothreonine- (pThr) containing peptides with the chemical groups that are reactive to biotin-containing tags (e.g., phosphoprotein isotope-coded affinity tag or PhIAT)[30-32] or solid-phase tags (e.g., phosphoprotein isotope-coded solid-phase tag or PhIST).[33,34] The tagged peptides can be captured through affinity purification or released from the solid-phase beads through either photo- or acid-cleavage. These steps can be combined and simplified in the method named Solid-phase Michael Addition (SMA) where tagging and capture of phosphopeptides can be performed in one single reaction vessel.[35] Alternatively, Knight et al used cysteamine during Michael addition to convert pSer and pThr residues into lysine analogs, which can be cleaved by trypsin or Lys-C proteases.[36] When a solid-phase cysteamine equivalent is used, this method is amenable to phosphopeptide enrichment. Unfortunately, the chemistry involved in the above procedures is not amenable to phosphotyrosine (pTyr) residues. In addition, the hydroxide-mediated β-elimination not only reacts with pSer and pThr but also is accessible to the O-glycosylated Ser and Thr residues, albeit with relatively lower reactivity. Consequently, sometimes it is difficult to distinguish between phosphorylation and glycosylation sites unambiguously if a sample has both types of modifications.

Additional chemical approaches relying on a phosphoramidate condensation reaction for the isolation of phosphopeptides have also been developed. The solid-phase phosphopeptide capture-and-release approach developed by Zhou et al includes six steps with the key procedures being carbodiimide-catalyzed condensation reaction for the formation of phosphoramidate bonds and solid-phase capture.[37] This approach is equally applicable to pSer-, pThr- and pTyr-containing peptides. The method was further simplified by Tao et al using a soluble dendrimer (a synthetic polyamine) as a source of amino groups in the condensation reaction.[38] Phosphopeptides are covalently linked via phosphoramidate bonds to the dendrimer which is readily isolated using filters or size-exclusion chromatography. The original phosphopeptides can be recovered by cleaving the phosphoramidate via acid hydrolysis. The advantage of this approach is that the soluble dendrimer supports a homogeneous reaction with adequate amino groups, allowing higher yield of phosphopeptides using this immobilization method than solid-phase support.[38] Nonetheless, the drawbacks are that the condensation reaction between phosphate and amino groups is extremely slow and the carboxylate groups have to be blocked (e.g., with methylation) prior to dendrimer conjugation chemistry. The overall phosphopeptide recovery of this approach is only 35%. However, a recent study showed that 535 phosphorylation sites from *Drosophila melanogaster* Kc167 cells were

Table 1. A comparison of phosphoprotein/phosphopeptide enrichment technologies for mass spectrometry analysis

Strategy	Method[a]	Targeted Residue[b]	Specificity[c]	Recovery	Side Reactions and Comments
Chemical modification	β-elimination and Michael addition (PhIAT, PhIST, SMA, etc.)	pS, pT	~70% (PhIST), 51% (SMA)	~60% (PhIAT), ~80% (PhIST)	Potentially capture O-linked glycoproteins and lipoproteins; reaction with Cys residues (protection needed); potential protein degradation; potential biased reaction toward pS compared to pT.
	Phosphoramidate condensation reaction	pS, pT, pY	>80% (Zhou, 2001)	~20% (Zhou, 2001), >35% (dendrimer)	The reaction is slow; carboxylate group protection needed; amino protection needed (Zhou, 2001).
Immunoaffinity	pY antibodies	pY	NA	~10-50% (Rush, 2005)	No side reactions.
	pS/pT antibodies	pS, pT	NA	NA	No side reactions; not all anti-pS/pT antibodies are applicable.
Ion exchange chromatography	SCX	pS, pT, pY	<40% (Trinidad, 2006)	NA	Phosphopeptides with a charge state other than 1+ are not amenable.
Metal ion affinity	IMAC (Fe^{3+}, Ga^{3+}, etc.)	pS, pT, pY	30-76% (without methylation); 85-95% (with methylation or prefractionation)	60-77%	Deamidation of Gln and Asn during methylation reaction; a bias toward multiply phosphorylated peptides when a low capacity IMAC column is used.
	TiO_2	pS, pT, pY	~40-50% (without methylation); ~99% (with methylation or prefractionation)	70-90%	No side reactions if phosphopeptides are not methylated.

a) Major enrichment methods are listed. PhIAT, phosphoprotein isotope-coded affinity tag; PhIST, phosphoprotein isotope-coded solid-phase tag; SMA, solid-phase Michael addition; SCX, strong cation exchange; IMAC, immobilized metal affinity chromatography; TiO_2, titanium dioxide. b) Considering three major phosphorylated residues: phosphoserine (pS), phosphothreonine (pT) and phosphotyrosine (pY). c) Specificity is based on the results from complex peptide/protein mixtures (i.e., proteome). NA, not available. Modified with permission from: Yu LR, Issaq HJ, Veenstra TD. Phosphoproteomics for the discovery of kinases as cancer biomarkers and drug targets. Proteomics Clin Appl 2007; 1:1042-1057; ©Wiley-VCH Verlag GmbH & Co. KGaA.

identified using a modified phosphoramidate chemistry-based enrichment,[39] so there are some promising uses of this approach.

Most of the chemical enrichment methods result in a limited number of phosphoprotein identifications owing to the complexity and potential side reactions of chemical derivatization. Immunoprecipitation is an alternative to enrich phosphoproteins or phosphopeptides.[40-44] The approach is mainly suitable for Tyr phosphorylation analysis because of the availability of robust anti-pTyr antibodies with high specificity. A recent Tyr phosphorylation analysis of cancer cells using anti-pTyr antibodies resulted in the identification of approximately 200 pTyr sites from a single cell line in a single analysis.[41] As for anti-pSer and anti-pThr antibodies, the phosphoamino acid is part of the recognition epitope but is often not the major recognition factor between the antibody and the phosphoprotein. Therefore, a single anti-pSer or anti-pThr antibody may not achieve a broad coverage of pSer or pThr phosphoproteome. In addition, Gronborg et al found that only half of the tested anti-pSer and anti-pThr antibodies could be used for immunoprecipitation of proteins phosphorylated at Ser/Thr residues.[45]

Strong cation exchange chromatography (SCX) was employed by Gygi's lab to enrich for phosphopeptides.[28,46] The early eluted fractions, at pH 2.7, should be enriched with +1 or less charged tryptic peptides and many of which should be phosphopeptides because of the negative charge of phosphate. Non-phosphorylated peptides, having at least two positive charges, would elute later in SCX separation. In the analysis of a HeLa cell nuclear extract, 2002 phosphorylation sites from 967 proteins were determined using tandem MS in combination with this enrichment approach.[28] Unfortunately, many phosphopeptides are eluted in the later SCX fractions as well,[47] leading to the loss of a fair number of phosphopeptides with multiple net positive charges.

Recently, metal ion affinity chromatography has been explored and widely applied to the enrichment of phosphopeptides. The technique, Immobilized Metal Affinity Chromatography (IMAC), can be tracked back to the year 1975[48] and has been improved to enrich phosphopeptides.[49,50] IMAC resins are activated by metal ions such as Fe^{3+}, Ga^{3+}, etc. Elution of phosphopeptides is typically conducted using high pH or a phosphate buffer as well as acidic solutions.[51,52] A more recent study indicated that phosphopeptide recovery was increased when the elution buffer was at low pH (5% TFA and 45% acetonitrile).[53] IMAC takes advantage of the strong affinity between metal ions and phosphate groups; however the carboxylate groups in acidic amino acids (i.e., Glu and Asp) and the C-termini of peptides significantly contribute to nonspecific bindings. Blocking carboxylate groups by methyl esterification can obviously reduce nonspecific bindings;[50] however this step introduces additional complexity including incomplete esterification and deamidation of Asn and Gln residues.[47,54,55] Sample prefractionation such as using hydrophilic interaction chromatography (HILIC) prior to IMAC could dramatically improve the selectivity of phosphopeptide enrichment (≥99%).[56] A recent study from Gygi's lab identified 5635 nonredundant phosphorylation sites on 2328 proteins from the mouse liver by the SCX-IMAC tandem phosphopeptide enrichment strategy.[57] To achieve better coverage of a phosphoproteome, sequential elution from IMAC (SIMAC) can be used to isolate mono- and multi-phosphorylated peptides sequentially on the same IMAC column using acidic and basic solvents. SIMAC may minimize the IMAC bias toward multi-phosphorylated peptides and double the number of identified phosphorylation sites.[58]

More recently, titanium dioxide (TiO_2) has been demonstrated to possess high affinity for phosphopeptides.[55,59-61] However, significant nonspecific binding is an issue when proteome samples (without methylation) are applied to TiO_2 columns.[62] These nonspecific bindings can be minimized by conditioning and/or washing the TiO_2 column after sample loading with a buffer containing 2,5-dihydroxybenzoic acid (DHB),[55] aliphatic hydroxy acids,[63] or ammonium glutamate.[62] Washing the column after sample loading using an ammonium glutamate buffer dramatically reduced nonphosphopeptide binding to TiO_2, which resulted in the identification of more than 1000 distinct phosphosites from more than 800 HeLa cell phosphopeptides by one-dimensional LC-MS/MS-MS/MS/MS analysis.[62] In combination of SCX fractionation with TiO_2 enrichment, Olsen et al detected 6600 phosphorylation sites on 2244 proteins from the same HeLa cell line.[64] Compared to IMAC, much time can be saved using TiO_2 phosphopeptide enrichment, in which column activation using metal ions as in IMAC is avoided. TiO_2 is also more compatible than IMAC to many reagents, including salts, detergents and other small molecules.[65] Other metal affinity approaches using zirconium dioxide,[66,67] alumina[68,69] and Fe_3O_4/TiO_2 core/shell nanoparticles[70,71] for selective enrichment of phosphopeptides have also been reported. Although these metal oxide/hydroxide affinity approaches for phosphopeptide enrichment have been just developed or are still in development, they represent very promising new approaches for selective enrichment of phosphopeptides.

Phosphoprotein Identification by Mass Spectrometry

Two general approaches for phosphoprotein identification, top-down and bottom-up, have been applied in the field. In the top-down approach, intact proteins are analyzed. High-resolution mass spectrometers such as Fourier transform ion cyclotron resonance MS (FTICR-MS) are required in this approach to achieve high mass accuracy. Currently, this approach is mainly for the analysis of single proteins or simple protein mixtures.[72,73] There are only limited applications in online HPLC separation coupled with MS for the analysis of intact proteins and only a few tens or hundreds of unmodified intact proteins could be identified using this approach.[74,75] In contrast, the bottom-up approach is more flexible and allows the use of a wide spectrum of MS instruments. Two ionization methods, matrix-assisted laser desorption ionization (MALDI) and electrospray ionization (ESI), have been applied to generate phosphopeptide ions for MS analysis.[8] However, ESI is more readily compatible to online LC separations; hence its use is more prevalent for the analysis of complex phosphopeptide mixtures isolated from cell lysates, tissues, etc.

Tandem MS (MS/MS) is usually used for the identification of phosphopeptides. While pTyr residues are relatively stable during collision-induced dissociation (CID), the phosphate moieties of pSer and pThr residues are labile, resulting in significant loss of phosphoric acid upon collision activation. In tandem MS, precursor ion scanning of m/z -79 (PO_3^-) in negative ion mode and of m/z 216.043 in positive ion mode (the immonium ion of pTyr) has been developed for selective detection of phosphopeptides;[76,77] however effective identification of phosphopeptide sequences typically require an additional tandem MS in the positive ion mode.[78] Neutral loss scan can be performed to detect pSer/pThr phosphopeptides by monitoring the neutral loss of phosphoric acid (−98 Da).[79] Owing to a significant neutral loss of phosphate during CID for the pSer/pThr phosphopeptides, the MS/MS spectrum is often dominated by a neutral-loss molecular ion with very low intensities of fragment ions (Fig. 2B), thus unambiguous identification of phosphosites is sometimes difficult. Recently, data-dependent neutral-loss MS/MS/MS (MS^3) was developed for improved phosphopeptide identification.[28,62,80,81] In MS^3, the neutral-loss peptide ions in the MS/MS (MS^2) are selected for CID, generating fragment ions to facilitate phosphopeptide identification and phosphosite localization (Fig. 2A); however the method is mainly effective for monophosphopeptides. Cross-validation of MS^2 and MS^3 scans within each analysis provides extremely high confidence (>99.5%) in phosphopeptide identification and phosphosite localization.[62] However, only 20-30% of the phosphopeptides identified from MS^2 scans can be cross-validated by MS^3 scans. Phosphopeptide identification could be improved using "Pseudo MS^n", in which a composite spectrum was recorded for the fragment ions from both original precursor activation and all subsequent neutral loss product activations.[82]

Recently, electron capture dissociation (ECD) was developed for effective peptide backbone fragmentation.[83,84] ECD generates c- and z-types of peptide fragment ions without evident loss of phosphate groups from phosphopeptides and thus is a powerful tool for phosphosite determination.[83] It has great potential for the characterization of intact phosphoproteins as well.[84] However, ECD is currently limited to FTICR-MS. Sweet et al paired FTICR-ECD with linear ion trap CID scan events to improve the efficiency of phosphopeptide identification, which resulted in the identification of 906 distinct phosphopeptides from mouse fibroblast NIH 3T3 cells.[85] In the meantime, electron transfer dissociation (ETD) was also developed and implemented on a quadrupole linear ion trap mass spectrometer to generate fragment ions similar to those from ECD.[86] An example of ETD MS/MS spectrum of a phosphopeptide is shown in Figure 3. In the hybrid

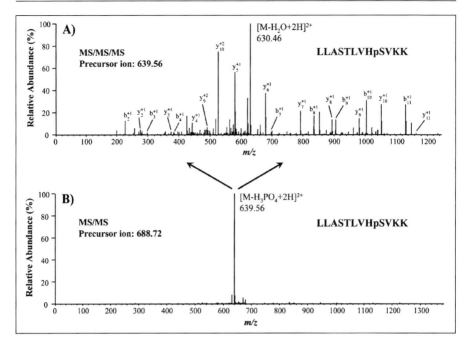

Figure 2. Tandem mass spectrometry analysis of phosphopeptides by collision-induced dissociation (CID). The doubly charged monophosphopeptide LLASTLVHpSVKK from Bcl-2-associated transcription factor 1 is subjected to CID-MS/MS (MS²) (B) and neutral loss-dependent MS/MS/MS (MS³) (A). The MS² spectrum (B) is dominated by the neutral-loss molecular ions (m/z 639.56) as a result of loss of phosphoric acid during CID from the peptide molecular ions (m/z 688.72) while all the other fragment ions are in very low abundance levels. Better fragmentation is shown in the MS³ spectrum (A) as a result the second stage CID of the isolated neutral-loss peptide ions (m/z 639.56) generated from MS². (Modified from Fig. 6, ref. 62.)

linear ion trap-Orbitrap (LTQ-Orbitrap) MS, ETD ions can be measured in the high-resolution Orbitrap analyzer. An ETD phosphoproteome study conducted in Hunt's lab demonstrated that 1252 phosphosites on 629 proteins were identified in a single experiment from enriched phosphopeptides within 30 μg of total yeast protein.[87] Similarly, Molina et al identified 1435 phosphorylation sites from human embryonic kidney 293 T-cells using the ETD fragmentation technique and found it identified 60% more phosphopeptides than CID, with an average of 40% more fragment ions that facilitated localization of phosphorylation sites.[88]

However, good performance of ETD or ECD fragmentation is limited to the peptides with high (≥+3) charge states.[89] With potential charge reduction by phosphorylation, the utility of ETD techniques in phosphoproteomics is restrained when tryptic peptides (mostly doubly charged) are analyzed. To improve the performance, alternative protein digestion methods, including the use of Lys-C, Lys-N and Glu-C proteinases or microwave D-cleavage digestion, can be used to generate relatively big peptides with multiple basic residues within a peptide, which increase the charges during ESI and facilitate ETD.[90,91] Adding m-nitrobenzyl alcohol (m-NBA) to the HPLC mobile phase has been reported to increase the charges of tryptic peptides and phosphopeptides in ESI.[92] On the other hand, ETD efficiency for doubly protonated peptides can be improved by a supplemental low-energy collisional activation (CAD) of the nondissociated (intact) electron-transfer (ET) product species ([M+2H]⁺·) (ETcaD), which leads to the near-exclusive generation of c- and z-type fragment ions with relatively high efficiency (77 ± 8%). ETcaD also demonstrated a significant improvement in protein sequence coverage compared to ETD or CAD

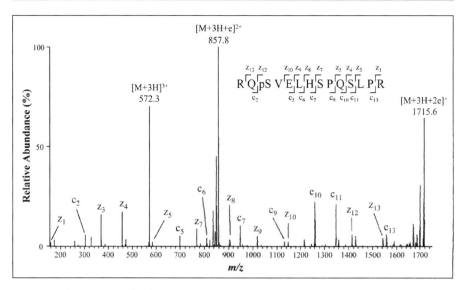

Figure 3. Electron transfer dissociation (ETD) tandem mass spectrometry analysis of the Aquaporin-2 phosphopeptide RQpSVELHSPQSLPR. Shown are c- and z-types of fragment ions generated during ETD without obvious neutral loss of phosphoric acid. Peptide molecular ions acquired electron(s) but not subjected to fragmentation are annotated.

alone.[93] The same strategy was employed by Domon et al to examine the *Drosophila melanogaster* phosphoproteome and reported that the method was especially beneficial for the characterization of large phosphopeptides including those with multiple phosphorylation sites.[94] In addition, peptide identification rates can be increased using a data-dependent decision tree algorithm in which real-time acquisition of a CID or ETD spectrum in a single analysis is dependent on precursor charge state and m/z.[89]

While phosphopeptide identification is usually achieved through a protein database search of experimental tandem MS data, the combinatorial expansion in search space with phosphorylation as a dynamic modification would potentially increase false positive identifications. High mass measurement accuracy obtained from high-resolution mass spectrometers is important to reduce the false positive rate. A recent study by Haas et al[95] demonstrated that a dramatic improvement of mass accuracy from 4.99±2.42 ppm to −0.25±1.46 ppm could be achieved by recalibrating the initial LTQ-FT MS data using polydimethylcyclosiloxane (PCM) ions as "lock masses", which were commonly detected as background ions in microcapillary LC-MS experiments. For low-quality MS/MS spectra, 90% more peptides were identified in combination with "lock mass"-calibrated FT MS spectra as compared with the linear ion trap MS spectra. In addition, more than a two-fold increase in phosphopeptide identifications was observed for the MS spectra acquired on the FT MS compared to those from linear ion trap MS. Similarly, an average absolute mass deviation of 0.48 ppm (standard deviation 0.38 ppm) and maximal deviation of less than 2 ppm were achieved by Olsen et al[96] for the MS and MS/MS spectra acquired with PCM "lock mass" ions on the LTQ-Orbitrap MS. With such high accuracy in MS and MS/MS measures, only one peptide sequence was assigned for most spectra after a database search. Therefore, mass measurement with high accuracy dramatically increases the confidence of peptide identification and is particular important for phosphopeptide analysis.

It is critical to validate tentatively identified phosphopeptides and phosphosites since a spectral match-based database search always has the potential of identifying false positive peptides and modification sites. While manual validation is tedious, computational algorithms greatly increase the speed of this process. An algorithm to calculate the ambiguity score (Ascore) was developed to estimate the probability of precise phosphorylation site assignment.[97] An Ascore of 19 or more indicates >99% certainty for correct phosphorylation site localization.[97] Another algorithm was developed by Olsen et al to calculate the posttranslational modification (PTM) probability score for phosphosite localization based on distinguishable fragment ions.[64] The method tests all possible combinations of phosphorylation sites in a peptide and the highest PTM score is assigned to the site considered to be correct. More recently, Lu et al developed two automatic validation methods for pSer/pThr-containing peptides, statistical multiple testing and support vector machine (SVM, a machine learning algorithm) binary classifier.[98] The methods conform very well with manual expert validation in a blinded test. However, these two approaches only assess whether the initially identified peptides are likely to be phosphorylated without considering the validity of phosphosites.

Global Quantitative Phosphoproteomic Technologies for Biomarker Discovery

To discover candidate phosphoprotein biomarkers, phosphoproteome analysis should be conducted in a quantitative manner. Label-free approaches for relative quantitation of proteome changes have been applied to some biological systems.[99-101] This approach allows a large number of samples to be compared but sample preparation and LC-MS conditions need to be strictly controlled. Accurate comparison of protein abundance changes can be achieved using stable isotopic labeling of proteins or peptides. To quantify protein phosphorylation absolutely,[102] it is necessary to synthesize stable isotope-coded standards as internal references.

Several approaches have been developed for stable isotope labeling of proteins. One of the commonly used labeling approaches is metabolic labeling, which is performed prior to protein extraction (pre-extraction). This type of labeling is usually achieved during cell culturing using $^{14}N/^{15}N$ enriched media[103] or stable isotope-coded amino acids.[104-107] It is preferable to use elements such as nitrogen and carbon because light and heavy versions of the labeled peptides are basically co-eluted from the reversed-phase LC column. A typical example of this type of labeling is named stable isotope labeling by amino acids in cell culture (SILAC)[107] in which the control and test cells are labeled with isotopically different versions (light and heavy) of amino acids such as Lys and/or Arg. A schematic approach of quantitative phosphoproteome analysis using SILAC and phosphopeptide enrichment is shown in Figure 4. Metabolic labeling allows the compared samples to be pooled at an early stage of analysis to avoid the variations produced in parallel processing of multiple samples; however the technique is only amenable to organisms and cells that can be metabolically labeled. Using this approach in combination of SCX fractionation and high-accuracy MS, Olsen et al quantitated dynamic phosphorylation changes of 1046 phosphopeptides (14% of ~6000 measured phosphosites) associated with EGF stimulation in HeLa cells.[64] The data revealed that different sites on the same protein are differentially regulated by phosphorylation and Tyr phosphorylation generally occurs earlier than Ser/Thr phosphorylation.[64] The dynamic phosphorylation data from this exemplified quantitative phosphoproteomic analysis were archived in the Phosida (*pho*sphorylation *si*te *da*tabase) database, providing a large resource for phosphoprotein biomarker discovery as well as cell signaling research for EGFR-amplified cancer.

Stable isotopic labeling can also be conducted after the protein is extracted (postextraction) if metabolic labeling is not possible. The labeling can be incorporated into the previously discussed β-elimination-based phosphopeptide enrichment methods for peptide quantitation, however hydrogen/deuterium (H/D) labeling is not recommended since this results in differential separation of H/D labeled peptides during LC. Trypsin-catalyzed $^{16}O/^{18}O$ exchange is another labeling strategy, in which two oxygen atoms of the C-terminal carboxylate group of each peptide can be replaced by ^{18}O atoms in ^{18}O-water via a trypsin-catalyzed oxygen exchange mechanism.[108-111] Other proteases could also be used for this labeling method. Since the resulting mass difference is only 4 Da for this type of labeling, high-resolution mass spectrometers are required and quantitation correction has to be performed in some cases because of the partial overlap between the isotope envelopes of the $^{16}O/^{18}O$-labeled peptide pair. In addition, isotope-coded amine-reactive reagents, for example the isobaric multiplexing tagging reagents

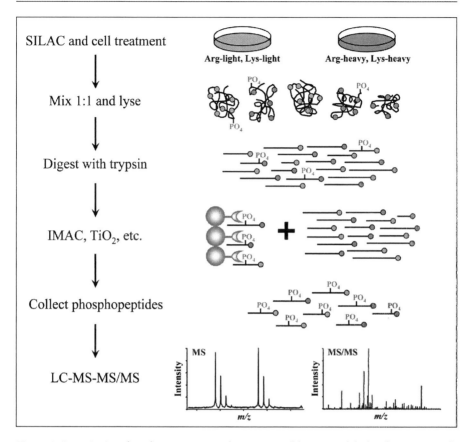

Figure 4. Quantitative phosphoproteome analysis using stable isotope labeling by amino acids in cell culture (SILAC) and mass spectrometry (MS). Two populations of cells are cultured in normal and stable isotope-coded arginine and lysine amino acids (Arg-light and Lys-light versus Arg-heavy and Lys-heavy). One population of cells is then stimulated and the other serves as a control. Cells are then mixed 1:1, lysed and digested with trypsin. Phosphopeptides are enriched using IMAC or TiO_2, etc. and then analyzed by liquid chromatography coupled on-line with tandem mass spectrometry (LC-MS/MS) to identify the exact sequences and phosphorylated sites of the enriched phosphopeptides. The MS scans are used to reconstruct the chromatographic peaks of the pair of phosphopeptides and relative quantitation is performed by comparing the peak areas of the phosphopeptides.

for relative and absolute protein quantitation (iTRAQ),[112] are another prevalent choice to label peptides prior to phosphopeptide enrichment. Zhang et al combined iTRAQ labeling with anti-pTyr immunoprecipitation and IMAC phosphopeptide enrichment for temporal analysis of tyrosine phosphorylation in response to EGFR activation and found that dynamic modules in the EGFR signaling network were consistent with particular cellular processes.[42] The iTRAQ reagents were initially designed for quantitation by CID tandem mass spectrometry and better for the use of time-of-flight MS analyzers since the reporter ions are in the low mass range. This type of labeling is not well suited for ion trap analyzers to perform CID for the detection of reporter ions. However, the utility of iTRAQ has been explored using ETD that is incorporated into ion trap and Orbitrap mass spectrometers.[113-115] Furthermore, pulsed Q dissociation (PQD) of linear ion trap MS and higher energy C-trap dissociation (HCD) of Orbitrap MS have been employed to measure iTRAQ-labeled peptides and applied to the analysis of phosphoproteome.[116,117] Bantscheff et al used PQD on an LTQ-Orbitrap MS to quantitatively measure the kinase interaction profile of imatinib, a tyrosine kinase inhibitor, in K-562 cells. After carefully optimizing instrument parameters, including collision energy, activation Q, delay time, ion isolation width, number of microscans and trapped ions, they achieved accurate peptide quantification at the 100 amol level. Compared with the results obtained from CID on a five-year-old Q-TOF MS, PQD on LTQ-Orbitrap MS doubled the number of proteins quantified although both approaches could reach similar quantitation errors of 10-15%.[118] PQD was also combined with ETD for quantitative measurement of iTRAQ-labeled phosphopeptides in which ETD improved phosphopeptide identification while PQD provided better detection and quantitation of iTRAQ reporter ions.[119]

Targeted Phosphoproteomic Approaches for Phosphoprotein Biomarker Validation and Qualification

Once the candidate biomarkers are discovered, the next step will be clinical validation and qualification of these candidates. The process refers to the determination of whether they are fit for the intended purpose (i.e., linked to the clinical endpoint) and requires the test platform to be analytically validated. Both are key steps before the biomarkers can be accepted by regulatory agencies for intended clinical use. There are many challenges to validate biomarkers in clinical settings, including proper patient recruitment (for prospective studies), analysis of large numbers of clinical samples, simultaneous measurement of multiple biomarkers, high stringency in biomarker measurement, high specificity and sensitivity, etc. Many assays are currently available to measure specific protein biomarkers; for example, an enzyme-linked immunosorbent assay (ELISA). However, it would be very challenging and expensive to develop multiplexed ELISAs for the measurement of multiple or even hundreds of biomarkers in a single assay. Another approach being pursued is the use of tissue microarrays for the detection of proteins/biomarkers for tens to hundreds of tissue samples.[120] Obviously, detection of phosphoprotein biomarkers using ELISA and tissue microarray approaches requires phosphosite-specific antibodies. The need for specific antibodies adds cost and delays qualification and validation of newly discovered biomarkers.

One of the key features in biomarker validation and use is accurate quantification in a relatively high-throughput fashion. As noted by Carr, the current proteomic workflow and throughput used in the discovery stage permits the analysis of a small number of samples (on the order of 10-20) but a large number of proteins (e.g., several thousand). However, to verify and validate biomarkers requires the analysis of hundreds or even thousands of samples with the focus usually on a small number of proteins (e.g., 1-10).[7] Selected reaction monitoring (SRM) or multiple reaction monitoring (MRM, when multiple product ions are monitored) has great potential as an MS-based tool for protein biomarker qualification and validation due to its high selectivity, sensitivity, dynamic range and throughput. SRM/MRM is typically conducted in a triple quadrupole MS where the first quadrupole (Q1) serves as a filter to select a peptide with a predefined specific m/z value and selected peptide ions are transmitted to the second quadrupole (Q2) and subjected to CID. Ions of one or more specific fragments are filtered in the third quadrupole (Q3) and then detected (Fig. 5). With minimal sample preparation, Anderson et al used this technique and quantified plasma proteins at high hundreds of ng/mL range with in-run coefficients of variation (CVs) of 2-22% and a dynamic range of 4.5 orders of magnitude.[121] The limits of detection (LODs) and limits of quantification (LOQs) have been achieved at low ng/mL ranges in plasma when stable isotope dilution (SID) or immunoaffinity enrichment such as Stable Isotope Standards and Capture by Anti-Peptide Antibodies (SISCAPA) is applied.[122,123] More recently, MRM allowed detection and quantification of proteins at a concentration below 50 copies/cell in total *S. cerevisiae* digests.[124] An interlaboratory comparison of quantification precision and reproducibility of MRM for plasma proteins showed that the CVs ranged from 4-14% for the samples prepared on one site and grew to 10-50% for the samples prepared in each individual lab.[125] Although the overall performance of the current MRM technique for protein quantification is not as stringent as that currently required for assays to be approved by the US Food and Drug Administration, the technique has a

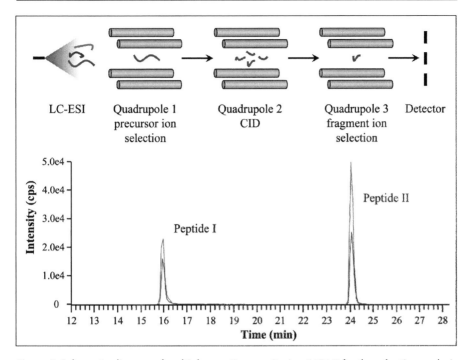

Figure 5. Schematic diagram of multiple reaction monitoring (MRM) for the selective analysis of peptides on a triple quadrupole mass spectrometer. In selected reaction monitoring (SRM) or MRM, the first quadrupole (Q1) serves as a filter to select a peptide with a predefined specific *m/z* value and selected peptide ions are transmitted to the second quadrupole (Q2) and subjected to fragmentation. Ions of one or more specific fragments are filtered in the third quadrupole (Q3) and then detected (upper panel). The lower panel is the result obtained from stable isotope dilution-multiple reaction monitoring (SID-MRM) quantitative assays of two targeted peptides. Shown are the extracted ion chromatograms (XIC) peaks of MRM Q1/Q3 ion pairs for natural (green) and heavy (red) isotope-coded peptides. A color version of this image is available at www.landesbioscience.com/curie.

great potential for candidate biomarker verification and to replace certain clinical immunoassays where interfering substances exist or multiplex measures are required.[125]

MRM has been applied to selective quantification of phosphopeptides and phosphorylation sites. Its early application to protein phosphorylation analysis demonstrated that MRM in conjunction with peptide sequencing, called multiple reaction monitoring-initiated detection and sequencing (MIDAS), was able to detect and determine phosphorylation sites at low femtomole levels of known proteins with high selectivity.[126] Heavy isotope labeled peptides or proteins can be applied as internal standards for MRM measurements to achieve absolute quantification of phosphorylation status of single proteins or protein mixtures. For example, Mayya et al studied the inhibitory multisite phosphorylation at Thr14 and Tyr15 of the cyclin-dependent kinases (CDKs) using phosphorylated and nonphosphorylated heavy isotope-labeled tryptic peptides as internal standards. They found that the transition to the mitotic phase was dominated by the conversion of "pT14-pY15" to the "T14-Y15" form of CDKs whereas the two monophosphorylated forms were considerably lower in abundance.[127] This kind of stable isotope dilution-multiple reaction monitoring mass spectrometry (SID-MRM-MS) has also been applied to absolute quantification of tyrosine phosphorylation on the focal adhesion kinase (FAK). Using isotopically labeled and phosphorylated FAK protein as the internal standard, Ciccimaro et al quantified phosphorylation of Y576 and Y577 within the activation loop domain of FAK as a result of interaction between

FAK and Src.[128] Application of MRM to global isotope labeling approaches such as iTRAQ has been demonstrated recently. This strategy enabled highly reproducible quantification of hundreds of nodes (phosphorylation sites) within a signaling network and across multiple conditions simultaneously as reported by Wolf-Yadlin et al.[129] They quantified temporal phosphorylation profiles of 222 tyrosine phosphorylated peptides across seven time points following EGF treatment, including 31 tyrosine phosphorylation sites not previously known to be regulated by EGF stimulation. The approach involved iTRAQ labeling, IMAC enrichment, MRM for *y*- and *b*-type ion transitions and iTRAQ ion transitions.[129] While iTRAQ is invaluable for MRM quantification, a new amine-specific nonisobaric variant of iTRAQ called mTRAQ (Applied Biosystems, Inc.) has been developed. Unlike iTRAQ, the same peptide labeled with all the three mTRAQ reagents is distinguishable in MS mode (three nonisobaric parent ions) and MS/MS mode (nonisobaric sequence ions). Therefore, the MRM transitions chosen for the labeled peptide are specific to each version which is different in precursor and product ion masses (for instances where the product ions retain the mTRAQ tags). In this way, the MRM quantification is performed using the nonisobaric sequence specific ions instead of reporter ions (in iTRAQ) to achieve higher specificity of measurement in complex mixtures. Using this method, DeSouza et al demonstrated the absolute quantification of pyruvate kinase-M1/M2 in samples of nonmalignant normal proliferative endometrium and endometrial cancer (Type I). The amounts of pyruvate kinase present in the endometrial cancer samples and nonmalignant controls were 85 nmol/g and 21-26 nmol/g of total protein, respectively and the results were confirmed by ELISA.[130]

Phosphoproteomics in Cancer Biomarker Discovery

While lots of efforts have been made to discover cancer biomarkers for many types of cancers, the majority of these efforts have focused on searching for unmodified proteins using MS-based quantitative proteomic technologies. Application of MS-based phosphoproteomic technologies to the identification of phosphoproteins including kinases as cancer biomarkers is an emerging field. Although many potential protein biomarkers have been identified from biofluids such as sera, the search for phosphoproteins in cerebrospinal fluid (CSF), serum and plasma using phosphoproteomic technologies is just beginning.[131-133] In the analysis of normal human CSF, 56 putative novel phosphorylation sites from 38 proteins were identified.[131] Using stringent filtering criteria, initial characterization of the serum phosphoproteome identified approximately 100 unique phosphopeptides and a lower than 1% false discovery rate (FDR).[132] In another study, 127 phosphosites in 138 phosphopeptides mapping to 70 phosphoproteins were identified from human plasma with a FDR < 1%.[133] Using two-dimensional polyacrylamide gel electrophoresis (2D-PAGE) and MS, Ogata et al found that the abundance level of phosphorylated fibrinogen-α-chain is elevated in the plasma of Stage III or IV ovarian cancer patients compared to that in normal controls.[134] Such studies provide evidence that phosphoproteins are present in normal biofluids and may be of potential use as noninvasive biomarkers in cancer patients.

Phosphoproteome-wide signaling studies can lead to the discovery of biomarkers useful for cancer diagnosis and/or for monitoring the response of drug treatment. Constitutive activation of the BCR-ABL kinase, resulted from fusion of a portion of the breakpoint cluster region (*BCR*) gene with c-*ABL* oncogene, is a major disease-causing factor for chronic myelogenous leukemia (CML).[135,136] Imatinib (Gleevec®, Novartis), a small-molecule inhibitor of ABL tyrosine kinase, has been approved to treat this type of disease. However, imatinib resistance arises in patients with advanced CML primarily because of point mutations in the kinase domain of BCR-ABL that tend to impair imatinib binding.[137] A study by Goss et al identified 188 Tyr-phosphorylated sites, of which 77 were novel, from 6 cell lines representing 3 BCR-ABL fusion types and 2 distinct cellular backgrounds using tyrosine phosphopeptide enrichment, SILAC and tandem MS.[138] Regardless of cellular background or fusion type, eight phosphoproteins (phosphosites) were consistently associated with activated BCR-ABL and could be defined as a BCR-ABL pTyr signature. Six of the eight phosphoproteins (phosphosites) were previously linked to BCR-ABL signaling: ABL (Y185, Y226 and Y393), BCR (Y177), Cbl (Y674), SHIP-2 (Y986 and Y1135), Shc (Y427) and VASP

(Y39). Two additional phosphoproteins previously unrelated to this signaling, CD2AP (Y548) and GRF1 (Y1106), were identified. In another study, a combination of 2D-PAGE and MS was used to investigate changes in the mitochondrial phosphoproteome in response to rapamycin, an mTOR inhibitor. The authors found this drug dramatically altered the mitochondrial phosphoproteome.[139] These studies demonstrate that phosphoproteomic approaches can be useful for discovering novel disease and drug response biomarkers.

Quantitative comparison of the phosphoproteomes from cells or tissues in normal and cancer states could uncover some unique "phosphosignatures" relevant to a disease state and serve as diagnostic markers. Tyrosine phosphoproteome comparison at the tissue level revealed that the patterns of Tyr-phosphorylated proteins varied markedly between different tissues, were very homogeneous in primary breast tumors and exhibited great diversity among liver cancers.[140] Comparing the Tyr phosphoproteomes between breast tumor tissues and the adjacent normal tissues, Lim et al identified breast tumor-specific tyrosine phosphoproteins, i.e., cytoskeletal proteins (actin, tubulin and vimentin) and chaperones (Hsp70, Hsc71 and Grp75).[140] By comparing benign, premalignant and tumor human breast epithelial cells and xenografts using phosphoproteomic approaches, Kim et al found that the abundance levels of many phosphoproteins and specific phosphorylation sites progressively increased in the cell lineage from benign to malignant cells with the greatest increase observed in breast tumor cells. Immunohistochemistry confirmed the levels of several phosphoproteins in xenografts and suggested that phospho-Akt and phospho-FOXO 1, 3a and 4 could be considered as biomarkers of tumorigenic risk.[141] Using TiO_2 phosphopeptide enrichment in combination with stable isotope labeling of cell cultures, solid tissue and plasma samples from healthy and diseased patients, Lee et al identified plectin-1 (pSer4253) and alpha-HS-glycoprotein (pSer138 and pSer312) as potential phospho-biomarkers for hepatocellular carcinoma.[142] The above studies suggest global quantitative phosphoproteome profiling is invaluable to the identification of cancer biomarkers.

Conclusion

The heterogeneous nature of cancer (e.g., differences in states of progression and between individuals) requires the discovery and qualification of more biomarkers for early detection of cancer, disease stratification and targeted cancer therapy. The advancements in global and targeted quantitative proteomics will continue to contribute to personalized cancer therapy in a large degree.[143] The advantage of identifying phosphoprotein biomarkers for cancer diagnosis is evident since phosphoproteins, such as kinases, play important roles in oncogenesis. Cancer-specific phosphosignatures will greatly facilitate therapeutic decision making, especially in the era of targeted and personalized cancer therapy. Specific phosphosignatures should aid in the prediction of therapeutic outcomes of a specific type or subtype of cancer and monitoring cancer treatment efficacy and, potentially, toxicity as well.

Techniques for phosphoprotein/phosphopeptide enrichment and subsequent MS analysis have been advancing rapidly. Nonetheless, there are still many challenges in the identification and quantification of phosphoproteins or phosphopeptides, among which include expansion of the dynamic range of phosphoproteome analysis, unambiguous characterization of phosphosites and absolute quantification of large numbers of phosphoproteins. Bioinformatics tools that aid in the validation of phosphopeptides at the mass spectral level are valuable in phosphoproteome analysis.[64,97,98] Phosphopeptide identification will be improved when multiple phosphopeptide fragmentation techniques and novel approaches are applied in the research field. MRM-based targeted phosphoproteomic approaches may play critical roles in obtaining detailed phosphorylation status of targeted proteins, especially for the further qualification of candidate phosphoprotein biomarkers already identified. The technique could be essential when large numbers of clinical samples are analyzed to verify phosphoprotein biomarkers for regulatory qualification and this type of analysis is being performed in a high throughput fashion. Since phosphoproteins could undergo dephosphorylation quickly when phosphatase inhibitors are not used during sample processing,[140] understanding and following the special requirements for clinical sample collection is required. It is expected that many more phosphoprotein cancer biomarkers will be identified and validated along with the continuous development and wide application of phosphoproteomic approaches to the analysis of large numbers of clinical cancer samples.

Acknowledgements

This work was supported with funds from National Center for Toxicological Research, U.S. Food and Drug Administration (NCTR/FDA) (L.R.Y.). The views presented in this chapter do not necessarily reflect those of the U.S. Food and Drug Administration.

References

1. Tabin CJ, Bradley SM, Bargmann CI et al. Mechanism of activation of a human oncogene. Nature 1982; 300(5888):143-149.
2. Weinberg RA. Tumor suppressor genes. Science 1991; 254(5035):1138-1146.
3. Wilhelm S, Carter C, Lynch M et al. Discovery and development of sorafenib: a multikinase inhibitor for treating cancer. Nat Rev Drug Discov 2006; 5(10):835-844.
4. Etzioni R, Urban N, Ramsey S et al. The case for early detection. Nat Rev Cancer 2003; 3(4):243-252.
5. Omenn GS. Strategies for plasma proteomic profiling of cancers. Proteomics 2006; 6(20):5662-5673.
6. Kuramitsu Y, Nakamura K. Proteomic analysis of cancer tissues: shedding light on carcinogenesis and possible biomarkers. Proteomics 2006; 6(20):5650-5661.
7. Rifai N, Gillette MA, Carr SA. Protein biomarker discovery and validation: the long and uncertain path to clinical utility. Nat Biotechnol 2006; 24(8):971-983.
8. Aebersold R, Mann M. Mass spectrometry-based proteomics. Nature 2003; 422(6928):198-207.
9. Phizicky E, Bastiaens PI, Zhu H et al. Protein analysis on a proteomic scale. Nature 2003; 422(6928):208-215.
10. Washburn MP, Wolters D, Yates JR, 3rd. Large-scale analysis of the yeast proteome by multidimensional protein identification technology. Nat Biotechnol 2001; 19(3):242-247.
11. Zhu H, Bilgin M, Bangham R et al. Global analysis of protein activities using proteome chips. Science 2001; 293(5537):2101-2105.
12. Cho WC. Contribution of oncoproteomics to cancer biomarker discovery. Mol Cancer 2007; 6:25.
13. Patwa TH, Zhao J, Anderson MA et al. Screening of glycosylation patterns in serum using natural glycoprotein microarrays and multi-lectin fluorescence detection. Anal Chem 2006; 78(18):6411-6421.
14. Yang Z, Harris LE, Palmer-Toy DE et al. Multilectin affinity chromatography for characterization of multiple glycoprotein biomarker candidates in serum from breast cancer patients. Clin Chem 2006; 52(10):1897-1905.
15. Salih E. Phosphoproteomics by mass spectrometry and classical protein chemistry approaches. Mass Spectrom Rev 2005; 24(6):828-846.
16. Vogelstein B, Kinzler KW. Cancer genes and the pathways they control. Nat Med 2004; 10(8):789-799.
17. Petricoin EF, Zoon KC, Kohn EC et al. Clinical proteomics: translating benchside promise into bedside reality. Nat Rev Drug Discov 2002; 1(9):683-695.
18. Jones PA, Baylin SB. The fundamental role of epigenetic events in cancer. Nat Rev Genet 2002; 3(6):415-428.
19. Hanahan D, Weinberg RA. The hallmarks of cancer. Cell 2000; 100(1):57-70.
20. Polanski M, Anderson NL. A list of candidate cancer biomarkers for targeted proteomics. Biomarker Insights 2006; 2:1-48.
21. Ludwig JA, Weinstein JN. Biomarkers in cancer staging, prognosis and treatment selection. Nat Rev Cancer 2005; 5(11):845-856.
22. Lim YP. Mining the tumor phosphoproteome for cancer markers. Clin Cancer Res 2005; 11(9):3163-3169.
23. Shawver LK, Slamon D, Ullrich A. Smart drugs: tyrosine kinase inhibitors in cancer therapy. Cancer Cell 2002; 1(2):117-123.
24. Slamon DJ, Clark GM, Wong SG et al. Human breast cancer: correlation of relapse and survival with amplification of the HER-2/neu oncogene. Science 1987; 235(4785):177-182.
25. Baker M. In biomarkers we trust? Nat Biotechnol 2005; 23(3):297-304.
26. Taguchi F, Solomon B, Gregorc V et al. Mass spectrometry to classify nonsmall-cell lung cancer patients for clinical outcome after treatment with epidermal growth factor receptor tyrosine kinase inhibitors: a multicohort cross-institutional study. J Natl Cancer Inst 2007; 99(11):838-846.
27. Boulay A, Zumstein-Mecker S, Stephan C et al. Antitumor efficacy of intermittent treatment schedules with the rapamycin derivative RAD001 correlates with prolonged inactivation of ribosomal protein S6 kinase 1 in peripheral blood mononuclear cells. Cancer Res 2004; 64(1):252-261.
28. Beausoleil SA, Jedrychowski M, Schwartz D et al. Large-scale characterization of HeLa cell nuclear phosphoproteins. Proc Natl Acad Sci USA 2004; 101(33):12130-12135.

29. Asano S, Park JE, Sakchaisri K et al. Concerted mechanism of Swe1/Wee1 regulation by multiple kinases in budding yeast. EMBO J 2005; 24(12):2194-2204.
30. Oda Y, Nagasu T, Chait BT. Enrichment analysis of phosphorylated proteins as a tool for probing the phosphoproteome. Nat Biotechnol 2001; 19(4):379-382.
31. Goshe MB, Conrads TP, Panisko EA et al. Phosphoprotein isotope-coded affinity tag approach for isolating and quantitating phosphopeptides in proteome-wide analyses. Anal Chem 2001; 73(11):2578-2586.
32. Goshe MB, Veenstra TD, Panisko EA et al. Phosphoprotein isotope-coded affinity tags: application to the enrichment and identification of low-abundance phosphoproteins. Anal Chem 2002; 74(3):607-616.
33. Qian WJ, Goshe MB, Camp DG, 2nd et al. Phosphoprotein isotope-coded solid-phase tag approach for enrichment and quantitative analysis of phosphopeptides from complex mixtures. Anal Chem 2003; 75(20):5441-5450.
34. Chowdhury SM, Munske GR, Siems WF et al. A new maleimide-bound acid-cleavable solid-support reagent for profiling phosphorylation. Rapid Commun. Mass Spectrom 2005; 19(7):899-909.
35. Tseng HC, Ovaa H, Wei NJ et al. Phosphoproteomic analysis with a solid-phase capture-release-tag approach. Chem Biol 2005; 12(7):769-777.
36. Knight ZA, Schilling B, Row RH et al. Phosphospecific proteolysis for mapping sites of protein phosphorylation. Nat Biotechnol 2003; 21(9):1047-1054.
37. Zhou H, Watts JD, Aebersold R. A systematic approach to the analysis of protein phosphorylation. Nat Biotechnol 2001; 19(4):375-378.
38. Tao WA, Wollscheid B, O'Brien R et al. Quantitative phosphoproteome analysis using a dendrimer conjugation chemistry and tandem mass spectrometry. Nat Methods 2005; 2(8):591-598.
39. Bodenmiller B, Mueller LN, Mueller M et al. Reproducible isolation of distinct, overlapping segments of the phosphoproteome. Nat Methods 2007; 4(3):231-237.
40. Salomon AR, Ficarro SB, Brill LM et al. Profiling of tyrosine phosphorylation pathways in human cells using mass spectrometry. Proc Natl Acad Sci USA 2003; 100(2):443-448.
41. Rush J, Moritz A, Lee KA et al. Immunoaffinity profiling of tyrosine phosphorylation in cancer cells. Nat Biotechnol 2005; 23(1):94-101.
42. Zhang Y, Wolf-Yadlin A, Ross PL et al. Time-resolved mass spectrometry of tyrosine phosphorylation sites in the epidermal growth factor receptor signaling network reveals dynamic modules. Mol Cell Proteomics 2005; 4(9):1240-1250.
43. Pandey A, Podtelejnikov AV, Blagoev B et al. Analysis of receptor signaling pathways by mass spectrometry: identification of vav-2 as a substrate of the epidermal and platelet-derived growth factor receptors. Proc Natl Acad Sci USA 2000; 97(1):179-184.
44. Blagoev B, Ong SE, Kratchmarova I et al. Temporal analysis of phosphotyrosine-dependent signaling networks by quantitative proteomics. Nat Biotechnol 2004; 22(9):1139-1145.
45. Gronborg M, Kristiansen TZ, Stensballe A et al. A mass spectrometry-based proteomic approach for identification of serine/threonine-phosphorylated proteins by enrichment with phospho-specific antibodies: identification of a novel protein, Frigg, as a protein kinase A substrate. Mol Cell Proteomics 2002; 1(7):517-527.
46. Ballif BA, Villen J, Beausoleil SA et al. Phosphoproteomic analysis of the developing mouse brain. Mol Cell Proteomics 2004; 3(11):1093-1101.
47. Trinidad JC, Specht CG, Thalhammer A et al. Comprehensive identification of phosphorylation sites in postsynaptic density preparations. Mol Cell Proteomics 2006; 5(5):914-922.
48. Porath J, Carlsson J, Olsson I et al. Metal chelate affinity chromatography, a new approach to protein fractionation. Nature 1975; 258(5536):598-599.
49. Andersson L, Porath J. Isolation of phosphoproteins by immobilized metal (Fe^{3+}) affinity chromatography. Anal Biochem 1986; 154(1):250-254.
50. Ficarro SB, McCleland ML, Stukenberg PT et al. Phosphoproteome analysis by mass spectrometry and its application to Saccharomyces cerevisiae. Nat Biotechnol 2002; 20(3):301-305.
51. Haydon CE, Eyers PA, Aveline-Wolf LD et al. Identification of novel phosphorylation sites on Xenopus laevis Aurora A and analysis of phosphopeptide enrichment by immobilized metal-affinity chromatography. Mol Cell Proteomics 2003; 2(10):1055-1067.
52. Liu H, Stupak J, Zheng J et al. Open tubular immobilized metal ion affinity chromatography combined with MALDI MS and MS/MS for identification of protein phosphorylation sites. Anal Chem 2004; 76(14):4223-4232.
53. Lee J, Xu Y, Chen Y et al. Mitochondrial phosphoproteome revealed by an improved IMAC method and MS/MS/MS. Mol Cell Proteomics 2007; 6:669-676.
54. Brill LM, Salomon AR, Ficarro SB et al. Robust phosphoproteomic profiling of tyrosine phosphorylation sites from human T-cells using immobilized metal affinity chromatography and tandem mass spectrometry. Anal Chem 2004; 76(10):2763-2772.

55. Larsen MR, Thingholm TE, Jensen ON et al. Highly selective enrichment of phosphorylated peptides from peptide mixtures using titanium dioxide microcolumns. Mol Cell Proteomics 2005; 4(7):873-886.
56. McNulty DE, Annan RS. Hydrophilic interaction chromatography reduces the complexity of the phosphoproteome and improves global phosphopeptide isolation and detection. Mol Cell Proteomics 2008; 7(5):971-980.
57. Villen J, Beausoleil SA, Gerber SA et al. Large-scale phosphorylation analysis of mouse liver. Proc Natl Acad Sci USA 2007; 104(5):1488-1493.
58. Thingholm TE, Jensen ON, Robinson PJ et al. SIMAC (sequential elution from IMAC), a phosphoproteomics strategy for the rapid separation of monophosphorylated from multiply phosphorylated peptides. Mol Cell Proteomics 2008; 7(4):661-671.
59. Pinkse MW, Uitto PM, Hilhorst MJ et al. Selective isolation at the femtomole level of phosphopeptides from proteolytic digests using 2D-NanoLC-ESI-MS/MS and titanium oxide precolumns. Anal Chem 2004; 76(14):3935-3943.
60. Kuroda I, Shintani Y, Motokawa M et al. Phosphopeptide-selective column-switching RP-HPLC with a titania precolumn. Anal Sci 2004; 20(9):1313-1319.
61. Sano A, Nakamura H. Titania as a chemo-affinity support for the column-switching HPLC analysis of phosphopeptides: application to the characterization of phosphorylation sites in proteins by combination with protease digestion and electrospray ionization mass spectrometry. Anal Sci 2004; 20(5):861-864.
62. Yu LR, Zhu Z, Chan KC et al. Improved titanium dioxide enrichment of phosphopeptides from HeLa cells and high confident phosphopeptide identification by cross-validation of MS/MS and MS/MS/MS spectra. J Proteome Res 2007; 6(11):4150-4162.
63. Sugiyama N, Masuda T, Shinoda K et al. Phosphopeptide enrichment by aliphatic hydroxy acid-modified metal oxide chromatography for nano-LC-MS/MS in proteomics applications. Mol Cell Proteomics 2007; 6(6):1103-1109.
64. Olsen JV, Blagoev B, Gnad F et al. Global, in vivo and site-specific phosphorylation dynamics in signaling networks. Cell 2006; 127(3):635-648.
65. Jensen SS, Larsen MR. Evaluation of the impact of some experimental procedures on different phosphopeptide enrichment techniques. Rapid Commun. Mass Spectrom 2007; 21(22):3635-3645.
66. Kweon HK, Hakansson K. Selective zirconium dioxide-based enrichment of phosphorylated peptides for mass spectrometric analysis. Anal Chem 2006; 78(6):1743-1749.
67. Zhou H, Xu S, Ye M et al. Zirconium phosphonate-modified porous silicon for highly specific capture of phosphopeptides and MALDI-TOF MS analysis. J Proteome Res 2006; 5(9):2431-2437.
68. Wolschin F, Wienkoop S, Weckwerth W. Enrichment of phosphorylated proteins and peptides from complex mixtures using metal oxide/hydroxide affinity chromatography (MOAC). Proteomics 2005; 5(17):4389-4397.
69. Chen CT, Chen WY, Tsai PJ et al. Rapid Enrichment of Phosphopeptides and Phosphoproteins from Complex Samples Using Magnetic Particles Coated with Alumina as the Concentrating Probes for MALDI MS Analysis. J Proteome Res 2007; 6(1):316-325.
70. Chen CT, Chen YC. Fe_3O_4/TiO_2 core/shell nanoparticles as affinity probes for the analysis of phosphopeptides using TiO2 surface-assisted laser desorption/ionization mass spectrometry. Anal Chem 2005; 77(18):5912-5919.
71. Li Y, Xu X, Qi D et al. Novel Fe_3O_4/TiO_2 core-shell microspheres for selective enrichment of phosphopeptides in phosphoproteome analysis. J Proteome Res 2008; 7(6):2526-2538.
72. Garcia BA, Pesavento JJ, Mizzen CA et al. Pervasive combinatorial modification of histone H3 in human cells. Nat Methods 2007; 4(6):487-489.
73. Ouvry-Patat SA, Torres MP, Quek HH et al. Free-flow electrophoresis for top-down proteomics by Fourier transform ion cyclotron resonance mass spectrometry. Proteomics 2008; 8(14):2798-2808.
74. Chi A, Bai DL, Geer LY et al. Analysis of intact proteins on a chromatographic time scale by electron transfer dissociation tandem mass spectrometry. Int J Mass Spectrom 2007; 259(1-3):197-203.
75. Bunger MK, Cargile BJ, Ngunjiri A et al. Automated proteomics of E. coli via top-down electron-transfer dissociation mass spectrometry. Anal Chem 2008; 80(5):1459-1467.
76. Mann M, Ong SE, Gronborg M et al. Analysis of protein phosphorylation using mass spectrometry: deciphering the phosphoproteome. Trends Biotechnol 2002; 20(6):261-268.
77. Steen H, Kuster B, Fernandez M et al. Detection of tyrosine phosphorylated peptides by precursor ion scanning quadrupole TOF mass spectrometry in positive ion mode. Anal Chem 2001; 73(7):1440-1448.
78. Old WM, Shabb JB, Houel S et al. Functional proteomics identifies targets of phosphorylation by B-Raf signaling in melanoma. Mol Cell 2009; 34(1):115-131.
79. Schlosser A, Pipkorn R, Bossemeyer D et al. Analysis of protein phosphorylation by a combination of elastase digestion and neutral loss tandem mass spectrometry. Anal Chem 2001; 73(2):170-176.
80. Gruhler A, Olsen JV, Mohammed S et al. Quantitative phosphoproteomics applied to the yeast pheromone signaling pathway. Mol Cell Proteomics 2005; 4(3):310-327.

81. Wolschin F, Lehmann U, Glinski M et al. An integrated strategy for identification and relative quantification of site-specific protein phosphorylation using liquid chromatography coupled to MS2/MS3. Rapid Commun. Mass Spectrom 2005; 19(24):3626-3632.
82. Schroeder MJ, Shabanowitz J, Schwartz JC et al. A Neutral Loss Activation Method for Improved Phosphopeptide Sequence Analysis by Quadrupole Ion Trap Mass Spectrometry. Anal Chem 2004; 76(13):3590-3598.
83. Stensballe A, Jensen ON, Olsen JV et al. Electron capture dissociation of singly and multiply phosphorylated peptides. Rapid Commun. Mass Spectrom 2000; 14(19):1793-1800.
84. Shi SD, Hemling ME, Carr SA et al. Phosphopeptide/phosphoprotein mapping by electron capture dissociation mass spectrometry. Anal Chem 2001; 73(1):19-22.
85. Sweet SM, Bailey CM, Cunningham DL et al. Large scale localization of protein phosphorylation by use of electron capture dissociation mass spectrometry. Mol Cell Proteomics 2009; 8(5):904-912.
86. Syka JE, Coon JJ, Schroeder MJ et al. Peptide and protein sequence analysis by electron transfer dissociation mass spectrometry. Proc Natl Acad Sci USA 2004; 101(26):9528-9533.
87. Chi A, Huttenhower C, Geer LY et al. Analysis of phosphorylation sites on proteins from Saccharomyces cerevisiae by electron transfer dissociation (ETD) mass spectrometry. Proc Natl Acad Sci USA 2007; 104(7):2193-2198.
88. Molina H, Horn DM, Tang N et al. Global proteomic profiling of phosphopeptides using electron transfer dissociation tandem mass spectrometry. Proc Natl Acad Sci USA 2007; 104(7):2199-2204.
89. Swaney DL, McAlister GC, Coon JJ. Decision tree-driven tandem mass spectrometry for shotgun proteomics. Nat Methods 2008; 5(11):959-964.
90. Hauser NJ, Han H, McLuckey SA et al. Electron transfer dissociation of peptides generated by microwave D-cleavage digestion of proteins. J Proteome Res 2008; 7(5):1867-1872.
91. Taouatas N, Drugan MM, Heck AJ et al. Straightforward ladder sequencing of peptides using a Lys-N metalloendopeptidase. Nat Methods 2008; 5(5):405-407.
92. Kjeldsen F, Giessing AM, Ingrell CR et al. Peptide sequencing and characterization of posttranslational modifications by enhanced ion-charging and liquid chromatography electron-transfer dissociation tandem mass spectrometry. Anal Chem 2007; 79(24):9243-9252.
93. Swaney DL, McAlister GC, Wirtala M et al. Supplemental activation method for high-efficiency electron-transfer dissociation of doubly protonated peptide precursors. Anal Chem 2007; 79(2):477-485.
94. Domon B, Bodenmiller B, Carapito C et al. Electron transfer dissociation in conjunction with collision activation to investigate the Drosophila melanogaster phosphoproteome. J Proteome Res 2009; 8(6):2633-2639.
95. Haas W, Faherty BK, Gerber SA et al. Optimization and use of peptide mass measurement accuracy in shotgun proteomics. Mol Cell Proteomics 2006; 5(7):1326-1337.
96. Olsen JV, de Godoy LM, Li G et al. Parts per million mass accuracy on an Orbitrap mass spectrometer via lock mass injection into a C-trap. Mol Cell Proteomics 2005; 4(12):2010-2021.
97. Beausoleil SA, Villen J, Gerber SA et al. A probability-based approach for high-throughput protein phosphorylation analysis and site localization. Nat Biotechnol 2006; 24(10):1285-1292.
98. Lu B, Ruse C, Xu T et al. Automatic validation of phosphopeptide identifications from tandem mass spectra. Anal Chem 2007; 79(4):1301-1310.
99. Wiener MC, Sachs JR, Deyanova EG et al. Differential mass spectrometry: a label-free LC-MS method for finding significant differences in complex peptide and protein mixtures. Anal Chem 2004; 76(20):6085-6096.
100. Qian WJ, Jacobs JM, Camp DG, 2nd et al. Comparative proteome analyses of human plasma following in vivo lipopolysaccharide administration using multidimensional separations coupled with tandem mass spectrometry. Proteomics 2005; 5(2):572-584.
101. Old WM, Meyer-Arendt K, Aveline-Wolf L et al. Comparison of label-free methods for quantifying human proteins by shotgun proteomics. Mol Cell Proteomics 2005; 4(10):1487-1502.
102. Gerber SA, Rush J, Stemman O et al. Absolute quantification of proteins and phosphoproteins from cell lysates by tandem MS. Proc Natl Acad Sci USA 2003; 100(12):6940-6945.
103. Conrads TP, Alving K, Veenstra TD et al. Quantitative analysis of bacterial and mammalian proteomes using a combination of cysteine affinity tags and 15N-metabolic labeling. Anal Chem 2001; 73(9):2132-2139.
104. Chen X, Smith LM, Bradbury EM. Site-specific mass tagging with stable isotopes in proteins for accurate and efficient protein identification. Anal Chem 2000; 72(6):1134-1143.
105. Zhu H, Pan S, Gu S et al. Amino acid residue specific stable isotope labeling for quantitative proteomics. Rapid Commun. Mass Spectrom 2002; 16(22):2115-2123.
106. Veenstra TD, Martinovic S, Anderson GA et al. Proteome analysis using selective incorporation of isotopically labeled amino acids. J Am Soc Mass Spectrom 2000; 11(1):78-82.

107. Ong SE, Blagoev B, Kratchmarova I et al. Stable isotope labeling by amino acids in cell culture, SILAC, as a simple and accurate approach to expression proteomics. Mol Cell Proteomics 2002; 1(5):376-386.
108. Schnolzer M, Jedrzejewski P, Lehmann WD. Protease-catalyzed incorporation of 18O into peptide fragments and its application for protein sequencing by electrospray and matrix-assisted laser desorption/ionization mass spectrometry. Electrophoresis 1996; 17(5):945-953.
109. Stewart, II, Thomson T, Figeys D. 18O labeling: a tool for proteomics. Rapid Commun. Mass Spectrom 2001; 15(24):2456-2465.
110. Wang YK, Ma Z, Quinn DF et al. Inverse 18O labeling mass spectrometry for the rapid identification of marker/target proteins. Anal Chem 2001; 73(15):3742-3750.
111. Yao X, Freas A, Ramirez J et al. Proteolytic 18O labeling for comparative proteomics: model studies with two serotypes of adenovirus. Anal Chem 2001; 73(13):2836-2842.
112. Ross PL, Huang YN, Marchese JN et al. Multiplexed protein quantitation in Saccharomyces cerevisiae using amine-reactive isobaric tagging reagents. Mol Cell Proteomics 2004; 3(12):1154-1169.
113. Han H, Pappin DJ, Ross PL et al. Electron transfer dissociation of iTRAQ labeled peptide ions. J Proteome Res 2008; 7(9):3643-3648.
114. Phanstiel D, Zhang Y, Marto JA et al. Peptide and protein quantification using iTRAQ with electron transfer dissociation. J Am Soc Mass Spectrom 2008; 19(9):1255-1262.
115. Phanstiel D, Unwin R, McAlister GC et al. Peptide quantification using 8-plex isobaric tags and electron transfer dissociation tandem mass spectrometry. Anal Chem 2009; 81(4):1693-1698.
116. Griffin TJ, Xie H, Bandhakavi S et al. iTRAQ reagent-based quantitative proteomic analysis on a linear ion trap mass spectrometer. J Proteome Res 2007; 6(11):4200-4209.
117. Boja ES, Phillips D, French SA et al. Quantitative mitochondrial phosphoproteomics using iTRAQ on an LTQ-Orbitrap with high energy collision dissociation. J Proteome Res 2009; 8(10):4665-4675.
118. Bantscheff M, Eberhard D, Abraham Y et al. Quantitative chemical proteomics reveals mechanisms of action of clinical ABL kinase inhibitors. Nat Biotechnol 2007; 25(9):1035-1044.
119. Yang F, Wu S, Stenoien DL et al. Combined pulsed-Q dissociation and electron transfer dissociation for identification and quantification of iTRAQ-labeled phosphopeptides. Anal Chem 2009; 81(10):4137-4143.
120. Kononen J, Bubendorf L, Kallioniemi A et al. Tissue microarrays for high-throughput molecular profiling of tumor specimens. Nat Med 1998; 4(7):844-847.
121. Anderson L, Hunter CL. Quantitative mass spectrometric multiple reaction monitoring assays for major plasma proteins. Mol Cell Proteomics 2006; 5(4):573-588.
122. Keshishian H, Addona T, Burgess M et al. Quantitative, multiplexed assays for low abundance proteins in plasma by targeted mass spectrometry and stable isotope dilution. Mol Cell Proteomics 2007; 6(12):2212-2229.
123. Anderson NL, Anderson NG, Haines LR et al. Mass spectrometric quantitation of peptides and proteins using Stable Isotope Standards and Capture by Anti-Peptide Antibodies (SISCAPA). J Proteome Res 2004; 3(2):235-244.
124. Picotti P, Bodenmiller B, Mueller LN et al. Full dynamic range proteome analysis of S. cerevisiae by targeted proteomics. Cell 2009; 138(4):795-806.
125. Addona TA, Abbatiello SE, Schilling B et al. Multi-site assessment of the precision and reproducibility of multiple reaction monitoring-based measurements of proteins in plasma. Nat Biotechnol 2009; 27(7):633-641.
126. Unwin RD, Griffiths JR, Leverentz MK et al. Multiple reaction monitoring to identify sites of protein phosphorylation with high sensitivity. Mol Cell Proteomics 2005; 4(8):1134-1144.
127. Mayya V, Rezual K, Wu L et al. Absolute quantification of multisite phosphorylation by selective reaction monitoring mass spectrometry: determination of inhibitory phosphorylation status of cyclin-dependent kinases. Mol Cell Proteomics 2006; 5(6):1146-1157.
128. Ciccimaro E, Hanks SK, Yu KH et al. Absolute quantification of phosphorylation on the kinase activation loop of cellular focal adhesion kinase by stable isotope dilution liquid chromatography/mass spectrometry. Anal Chem 2009; 81(9):3304-3313.
129. Wolf-Yadlin A, Hautaniemi S, Lauffenburger DA et al. Multiple reaction monitoring for robust quantitative proteomic analysis of cellular signaling networks. Proc Natl Acad Sci USA 2007; 104(14):5860-5865.
130. DeSouza LV, Taylor AM, Li W et al. Multiple reaction monitoring of mTRAQ-labeled peptides enables absolute quantification of endogenous levels of a potential cancer marker in cancerous and normal endometrial tissues. J Proteome Res 2008; 7(8):3525-3534.
131. Bahl JM, Jensen SS, Larsen MR et al. Characterization of the human cerebrospinal fluid phosphoproteome by titanium dioxide affinity chromatography and mass spectrometry. Anal Chem 2008; 80(16):6308-6316.
132. Zhou W, Ross MM, Tessitore A et al. An initial characterization of the serum phosphoproteome. J Proteome Res 2009; 8(12):5523-5531.

133. Carrascal M, Gay M, Ovelleiro D et al. Characterization of the human plasma phosphoproteome using linear ion trap mass spectrometry and multiple search engines. J Proteome Res 2010; 9(2):876-884.

134. Ogata Y, Hepplmann CJ, Charlesworth MC et al. Elevated levels of phosphorylated fibrinogen-alpha-isoforms and differential expression of other posttranslationally modified proteins in the plasma of ovarian cancer patients. J Proteome Res 2006; 5(12):3318-3325.

135. Ben-Neriah Y, Daley GQ, Mes-Masson AM et al. The chronic myelogenous leukemia-specific P210 protein is the product of the bcr/abl hybrid gene. Science 1986; 233(4760):212-214.

136. Deininger MW, Druker BJ. Specific targeted therapy of chronic myelogenous leukemia with imatinib. Pharmacol Rev 2003; 55(3):401-423.

137. Shah NP, Nicoll JM, Nagar B et al. Multiple BCR-ABL kinase domain mutations confer polyclonal resistance to the tyrosine kinase inhibitor imatinib (STI571) in chronic phase and blast crisis chronic myeloid leukemia. Cancer Cell 2002; 2(2):117-125.

138. Goss VL, Lee KA, Moritz A et al. A common phosphotyrosine signature for the Bcr-Abl kinase. Blood 2006; 107(12):4888-4897.

139. Schieke SM, Phillips D, McCoy JP, Jr. et al. The mammalian target of rapamycin (mTOR) pathway regulates mitochondrial oxygen consumption and oxidative capacity. J Biol Chem 2006; 281(37):27643-27652.

140. Lim YP, Wong CY, Ooi LL et al. Selective tyrosine hyperphosphorylation of cytoskeletal and stress proteins in primary human breast cancers: implications for adjuvant use of kinase-inhibitory drugs. Clin Cancer Res 2004; 10(12 Pt 1):3980-3987.

141. Kim SH, Miller FR, Tait L et al. Proteomic and phosphoproteomic alterations in benign, premalignant and tumor human breast epithelial cells and xenograft lesions: biomarkers of progression. Int J Cancer 2009; 124(12):2813-2828.

142. Lee HJ, Na K, Kwon MS et al. Quantitative analysis of phosphopeptides in search of the disease biomarker from the hepatocellular carcinoma specimen. Proteomics 2009; 9(12):3395-3408.

143. Koomen JM, Haura EB, Bepler G et al. Proteomic contributions to personalized cancer care. Mol Cell Proteomics 2008; 7(10):1780-1794.

Usefulness of Immunomics in Cancer Biomarker Discovery

Julie Hardouin* and Michel Caron

Abstract

Cancer remains one of the leading causes of death worldwide. The identification of new useful and specific biomarkers is necessary in order to detect it earlier. We used cancer immunomics to search and identify proteins of interest in the case of breast or colorectal cancers. Two approaches were developed at the laboratory: the top-down SERological Proteome Analysis (SERPA) and the bottom-up MAPPing (Multiple Affinity Protein Profiling) strategies. The first one relied on two-dimensional electrophoresis (2-DE), immunoblotting, image analysis and mass spectrometry. The second approach deals with the use of two-dimensional immuno-affinity chromatography, enzymatic digestion of the antigens and analysis by tandem mass spectrometry.

Introduction

Despite advances in diagnosis, cancer remains a major cause of mortality worldwide.[1] Currently no specific and sensitive markers exist for early diagnosing disease development. The proteome is the global representative of all biological processes that take place during cancer development. The identification of specific biomarker is a real challenge since more than one million of proteins are present. Therefore, new strategies have to be developed in order to improve the detection and the identification of proteins of interest.[2-6]

Tumors are thought to release many proteins into the blood. Therefore, the diagnosis of cancer based on serum profiling is a very interesting concept. There is evidence that the immune system is on guard against different threats, including tumours.[7,8] Uncontrolled malignant growth can thus be characterized by the presence of auto-antibodies that precede clinical findings by months and years.

For identifying useful auto-antibodies as specific biomarkers, new proteomic approaches—defined by the neologism term immunomics—have been more and more used.[9,10] A first possible way is the SERPA approach (for Serological Proteome Analysis).[11] This top-down method relies on the intact protein analysis. The proteins are separated by two-dimensional gel electrophoresis (2-DE) and a differential analysis is used to compare healthy and patient cohorts. For this purpose, two dimensional western blotting is used to reveal auto-antigens corresponding to the auto-antibodies find specifically in patient's sera.[12-15] A second strategy named MAPPing-IT (Multiple Affinity Protein Profiling—Ion Trap) was recently developed.[10,14,16] This bottom-up approach deals with the protein separation using multi-dimensional chromatography: affinity chromatography steps and reverse phase chromatography). Generally, for immunomics studies, the auto-antigens corresponding to patient's auto-antibodies are purified and then their complex mixture is digested and analyzed by liquid chromatography at nanoflow rate coupled to a tandem mass spectrometry study (nanoLC-MS/MS).

*Corresponding Author: Julie Hardouin—Université de Rouen, Laboratoire Polymères, Biopolymères, Surfaces, UMR CNRS 6270, équipe Biofilms, Résistance, Interactions, Cellules-Surface, 76821 Mont-Saint-Aignan cedex, France. Email: julie.hardouin@univ-rouen.fr

Omics Technologies in Cancer Biomarker Discovery, edited by Xuewu Zhang.
©2011 Landes Bioscience.

Serological Proteome Analysis

The term serological proteome analysis (SERPA) was proposed by Klade et al[11] in 2001 for a top-down usable method for immunomics. SERPA (also called PROTEOMEX by Siegler et al[17,18]) is based on a classical proteomics workflow. Protein extracts from cancer cells are used as source of antigens and are separated by 2-DE. Then, they are transferred onto a polyvinyldiene difluoride (PVDF) membrane. The blots are incubated with well characterized serum species coming from patients or sera from healthy volunteers.[11-13] The transferred proteins on these 2-D blots are stained with colloidal gold. The spots on the blots are then compared to a preparative gel. The spots of interest are also excised and digested by a specific enzyme (trypsin). The peptide mixtures are analyzed by mass spectrometry (MS) and identified after searching in the international databanks.[19,20] Figure 1 summarizes the experimental SERPA approach for the search and the identification of proteins recognized by patient's sera.

Our laboratory emphasized on an immunomics study of breast cancer.[12,13,15,21] Breast cancer is the most common malignancy among women.[1] We used the SERPA strategy for identifying breast cancer protein that gave a humoral response. First, data pointed out the occurrence of auto-antibodies of immunoglobulin G (IgG) isotype highly conserved between individuals. It is assumed that two-thirds of the IgG of healthy individuals may possess auto reactive affinities.[22] We observed that they were directed against a limited set of antigens independently of the cancer status. Secondly, a limited number of proteins react preferentially with cancer sera. In some

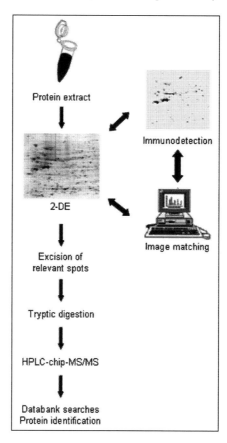

Figure 1. Description of the top-down SERPA strategy for the identification of cancer-associated auto-antigens.

cases, the patient's sera recognized preferentially isoforms of a same protein but having different isoelectric point. These isoforms corresponded to a same polypeptidic chain modified by different posttranslational modifications such as phosphorylation, methylation, acetylation, etc.[23,24] It was then considered that specific immune responses were induced in the patients by these modifications.

Multiple Affinity Protein Profiling

The term MAPPing-IT, for multiple affinity protein profiling—ion trap, has been proposed for a chromatography based method aiming at the identification of specific auto-antibodies through the purification and identification of complementary auto-antigens.

A preliminary step was the preparation of different immuno-affinity supports. For this purpose, IgG fractions were separated by affinity chromatography, using thiophilic adsorption chromatography, from sera of both healthy volunteers (controls) and patients.[25] These fractions were then coupled to agarose beads for the preparation of immuno-affinity supports.[25] A first type of support, used for the first dimension of MAPPing, was obtained by coupling a pool of IgGs from controls. For the second dimension of MAPPing, a series of supports were prepared by coupling individual fractions of IgG from patients, in order to obtain an immuno-affinity column for each patient.

Figure 2 describes the different steps (or dimensions) for the identification of antigens recognized by patients' auto-antibodies by MAPPing-IT. Proteins extracted from cells were separated by a two dimensional (2D) immuno-affinity chromatography. In order to isolate specifically, from

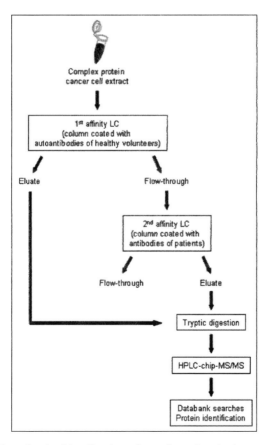

Figure 2. Methodology for the identification of proteins using the bottom-up MAPPing approach. (LC = liquid chromatography).

this complex proteome, the antigens recognized by the antibodies of patients, it was needed to simplify this proteome. This simplification was thus obtained during the first dimension on the first immuno-affinity support. This first dimension was very important since it allowed us to deplete the protein extract in antigens recognized by healthy volunteers' antibodies. The proteins that did not bind to this first column (flow through) were used for the second affinity chromatography step: the purification of antigens recognized by the antibodies of patients (eluate). Several columns could be used in parallel permitting the analyses in a same run on different columns, each of them being coated with the IgGs of a patient. The proteins (antigens) present in the eluates issued from this second step were then digested by the specific endoprotease trypsin and the mixtures of tryptic peptides issued from these antigens were analyzed by nanoLC coupled to MS/MS.[10,14,16] In our laboratory, we applied this strategy in the case of colorectal cancer in order to isolate and then to identify proteins recognized specifically by the antibodies of the patients.

Improved Identification of Protein/Biomarker Identification in Complex Mixtures Using Protein Signatures

When a protein has been identified and characterized as an exploratory biomarker, many experiments and correlation tests have to be done to validate this characterization, for instance on large populations of sera or plasma (healthy volunteers, patients, others pathologies, etc.). Generally, these experiments are complicated by the fact that these exploratory biomarkers correspond to low abundant proteins present in complex protein mixtures. In our approaches, the analysis of these complex mixtures were done using nanoLC-MS/MS. Classically, the identification of the protein components only focuses on the mass to charge ratio of the tryptic peptides issued form these components. In this situation, proteins are often identified with a limited number of peptides and sometimes with only one peptide.[26] Then, mass data is not always enough to validate without any doubt the presence of the peptide and then of the corresponding protein in the sample. An interesting approach is to define for each peptide a specific signature that can characterize this peptide with a better relevance than using only the mass to charge ratio. We choose to use data obtained during nanoLC-MS/MS analysis to create these signatures. At least, two data recorded

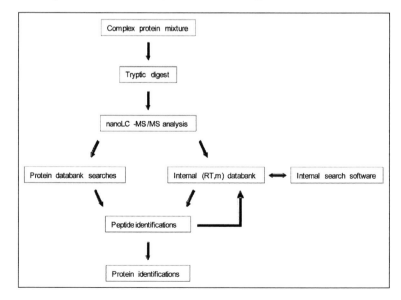

Figure 3. Principle of protein or biomarker identification using in a first step, the mass to charge ratio data for the search in the international protein databanks and in a second step, the (RT, m) signatures for the search in an internal databank.

during the nanoLC-MS/MS analysis can be used: the retention time (RT) easily available from the chromatogram and the mass to charge ratio (m) read on the mass spectrum.[27-30] The formation of these couples (RT; m) for some peptides issued from a protein/biomarker (here an antigen recognized specifically by the antibodies of patients) provides a helpful and specific signature of this exploratory biomarker.[31-33] A two dimensional map can be formed summarizing for each peptide its mass to charge ratio and the corresponding retention time. In order to measure reproducible retention time, we choose a microfluidic device named HPLC-chip-MS.[34-37] The chip integrates an enrichment column, a separation column and an electrospray tip. As less connections and capillaries are needed, more repeatable retention time measurements can be obtained than in classical HPLC systems. Furthermore, higher sensitivity is observed.[35] This technology was used to have repeatable RT measurements and then to create specific signatures (Fig. 3).[26,38] These signatures are then used to search the presence of specific peptides in complex mixtures from healthy volunteers, patients and patients having another disease in order to check the specificity of our protein of interest.

Acknowledgement

This work was supported by the French Ministère des Finances et de l'Industrie (NODDICCAP contract) and grants N° 3491 from the Association de la Recherche contre le Cancer, from the European Union's action in support of regional development (FEDER) and from bioMérieux, New Markers Department (Dr. Geneviève Choquet-Kastylevsky), Marcy-l'Etoile, France.

References

1. Jemal A, Tiwari RC, Murray T et al. Cancer statistics. CA Cancer J Clin 2004; 54:8-29.
2. Braziel RM, Shipp MA, Feldman AL et al. Molecular diagnostics. Hematology Am Soc Hematol Educ Program 2003; 279-293.
3. Caron M, Joubert-Caron R. Proteomics in hematologic malignancies. Expert Rev Proteomics 2005; 2:567-576.
4. Hanash S. Disease proteomics. Nature 2003; 422:226-232.
5. Joubert-Caron R, Caron M. Proteome analysis in the study of lymphoma cells. Mass Spectrom Rev 2005; 24:455-468.
6. Tyers M, Mann M. From genomics to proteomics. Nature 2003; 422:193-197.
7. Finn OJ. Immune response as a biomarker for cancer detection and a lot more. N Engl J Med 2005; 353:1288-1290.
8. Tan EM. Autoantibodies as reporters identifying aberrant cellular mechanisms in tumorigenesis. J Clin Invest 2001; 108:1411-1415.
9. Caron M, Choquet-Kastylevsky G, Joubert-Caron R. Cancer immunomics using autoantibody signatures for biomarker discovery. Mol Cell Proteomics 2007; 6:1115-1122.
10. Hardouin J, Lasserre JP, Sylvius L et al. Cancer immunomics: from serological proteome analysis to multiple affinity protein profiling. Ann N Y Acad Sci 2007; 1107:223-230.
11. Klade CS, Voss T, Krystek E et al. Identification of tumor antigens in renal cell carcinoma by serological proteome analysis. Proteomics 2001; 1:890-898.
12. Canelle L, Bousquet J, Pionneau C et al. An efficient proteomics-based approach for the screening of autoantibodies. J Immunol Methods 2005; 299:77-89.
13. Canelle L, Bousquet J, Pionneau C et al. A proteomic approach to investigate potential biomarkers directed against membrane-associated breast cancer proteins. Electrophoresis 2006; 27:1609-1616.
14. Caron M, Joubert-Caron, R, Canelle L et al. Serological proteome analysis (SERPA) and multiple affinity protein profiling (MAPPING) to discover cancer biomarkers. Mol Cell Proteomics 2005; 4:S142.
15. Pionneau C, Canelle L, Bousquet J et al. Proteomic analysis of membrane-associated proteins from the Breast cancer cell line MCF7. Cancer Gen Proteomics 2005; 2:199-208.
16. Hardouin J, Lasserre JP, Canelle L et al. Usefulness of autoantigens depletion to detect autoantibody signatures by multiple affinity protein profiling. J Sep Sci 2007; 30:352-358.
17. Seliger B, Kellner R. Design of proteome-based studies in combination with serology for the identification of biomarkers and novel targets. Proteomics 2002; 2:1641-1651.
18. Seliger B, Lichtenfels R, Kellner R. Detection of renal cell carcinoma-associated markers via proteome- and other 'ome'-based analyses. Brief Funct Genomic Proteomic 2003; 2:194-212.
19. Canelle L, Pionneau C, Marie A et al. Automating proteome analysis: improvements in throughput, quality and accuracy of protein identification by peptide mass fingerprinting. Rapid Commun Mass Spectrom 2004; 18:2785-2794.

20. Joubert-Caron R, Le Caer JP, Montandon F et al. Protein analysis by mass spectrometry and sequence database searching: a proteomic approach to identify human lymphoblastoid cell line proteins. Electrophoresis 2000; 21:2566-2575.

21. Hardouin J, Canelle L, Vlieghe C et al. Proteomic analysis of the MCF7 Breast cancer cell line. Cancer Gen Proteomics 2006; 3:355-368.

22. Scanlan MJ, Chen YT, Williamson B et al. Characterization of human colon cancer antigens recognized by autologous antibodies. Int J Cancer 1998; 76:652-658.

23. Imam-Sghiouar N, Joubert-Caron R, Caron M. Application of metal-chelate affinity chromatography to the study of the phosphoproteome. Amino Acids 2005; 28:105-109.

24. Imam-Sghiouar N, Laude-Lemaire I, Labas V et al. Subproteomics analysis of phosphorylated proteins: application to the study of B-lymphoblasts from a patient with Scott syndrome. Proteomics 2002; 2:828-838.

25. Hardouin J, Duchateau M, Canelle L et al. Thiophilic adsorption revisited. J Chromatogr B Analyt Technol Biomed Life Sci 2007; 845:226-231.

26. Hardouin J, Joubert-Caron R, Caron M. HPLC-chip-mass spectrometry for protein signature identifications. J Sep Sci 2007; 30:1482-1487.

27. Berg M, Parbel A, Pettersen H et al. Reproducibility of LC-MS-based protein identification. J Exp Bot 2006; 57:1509-1514.

28. Pasa-Tolic L, Masselon C, Barry RC et al. Proteomic analyses using an accurate mass and time tag strategy. Biotechniques 2004; 37:621-624, 626-633, 636 passim.

29. Smith RD, Anderson GA, Lipton MS et al. High-performance separations and mass spectrometric methods for high-throughput proteomics using accurate mass tags. Adv Protein Chem 2003; 65:85-131.

30. Zimmer JS, Monroe ME, Qian WJ et al. Advances in proteomics data analysis and display using an accurate mass and time tag approach. Mass Spectrom Rev 2006; 25:450-482.

31. Fang R, Elias DA, Monroe ME et al. Differential label-free quantitative proteomic analysis of Shewanella oneidensis cultured under aerobic and suboxic conditions by accurate mass and time tag approach. Mol Cell Proteomics 2006; 5:714-725.

32. Masselon CD, Kieffer-Jaquinod S, Brugiere S et al. Influence of mass resolution on species matching in accurate mass and retention time (AMT) tag proteomics experiments. Rapid Commun Mass Spectrom 2008; 22:986-992.

33. Norbeck AD, Monroe ME, Adkins JN et al. The utility of accurate mass and LC elution time information in the analysis of complex proteomes. J Am Soc Mass Spectrom 2005; 16:1239-1249.

34. Ghitun M, Bonneil E, Fortier MH et al. Integrated microfluidic devices with enhanced separation performance: application to phosphoproteome analyses of differentiated cell model systems. J Sep Sci 2006; 29:1539-1549.

35. Hardouin J, Duchateau M, Joubert-Caron R et al. Usefulness of an integrated microfluidic device (HPLC-Chip-MS) to enhance confidence in protein identification by proteomics. Rapid Commun Mass Spectrom 2006; 20:3236-3244.

36. Vollmer M, Horth P, Rozing G et al. Multi-dimensional HPLC/MS of the nucleolar proteome using HPLC-chip/MS. J Sep Sci 2006; 29:499-509.

37. Yin H, Killeen K, Brennen R et al. Microfluidic chip for peptide analysis with an integrated HPLC column, sample enrichment column and nanoelectrospray tip. Anal Chem 2005; 77:527-533.

38. Fortier MH, Bonneil E, Goodley P et al. Integrated microfluidic device for mass spectrometry-based proteomics and its application to biomarker discovery programs. Anal Chem 2005; 77:1631-1640.

CHAPTER 8

Glycoproteomics in Cancer Biomarker Discovery

Byung-Gyu Kim and Je-Yoel Cho*

Abstract

Glycoproteomics is one of the most promising fields that can lead us to discover new biomarkers in various types of diseases including cancers. Glycoproteomics is a branch of proteomics and identifies and characterizes the proteins which contain carbohydrates as a posttranslational modification. It studies not only the levels of glycoproteins in the samples but also the changes of glycosylation patterns on the glycoproteins in various diseases through MS (mass spectrometry)-based analysis. This chapter covers the major glycosylation process in proteins and the technologies used in the glycoprotein analysis and also highlights the application of the glycoproteome analysis with regard to the discovery of diagnostic or prognostic biomarkers for more accurate cancer diagnosis and therapy and for better understanding of cancer progression.

Introduction

In medicine, the simplest definition of a biomarker is a molecule that indicates an alteration in physiology from normal. A more practical definition of a biomarker can be a molecule that is detected in an organism as means to examine organ function or other aspects of health. Therefore, a 'biomarker' is 'a characteristic that is objectively measured and evaluated as an indicator of normal biological processes, pathogenic processes, or pharmacologic responses to a therapeutic intervention.' (FDA Guidance of Industry, Pharmacogenomic Data Submissions, March 2005).[1] Biomarkers are not only useful in early diagnosis of cancers but also provide important information in cancer therapy such as verification of cancer staging, response to therapy, guidance on therapy and clinical end points or surrogate end points.[2-4] Among biochemical features or facets, proteins or peptides which are produced in body organ system that can be used to measure the progress of disease or the effects of treatment, since they are responsible for most serious diseases.[5-7] Thus protein biomarkers present in blood, tissues, urine, hair, etc. are keys to identifying disease and determining the optimum course of therapy. Among them, "cancer biomarkers" have contributed greatly to our current understanding of the heterogeneous nature of specific cancers and have led to improvements in treatment outcomes.[8-11] Cancer is a class of diseases in which a group of cells display uncontrolled growth (division beyond the normal limits), invasion (intrusion on and destruction of adjacent tissues) and sometimes metastasis (spread to other locations in the body via lymph or blood). Cancer biomarker—based diagnostics, therefore, can be applied to establish disease predisposition, early detection, cancer staging, therapy selection, identifying whether or not a cancer is metastatic, therapy monitoring, assessing prognosis and advances in the adjuvant setting. Currently, a number of cancer therapies are carried out based on specific cancer biomarkers.[12-16] These biomarkers are called as companion biomarkers and are very useful for the selection of specific drugs to specific type of cancers with specific molecular alterations.

*Corresponding Author: Je-Yoel Cho—Department of Biochemistry, School of Dentistry, Kyungpook National University, 101 Dong In-Dong, Jung-Gu, Daegu, South Korea 700-422. Email: jeycho@knu.ac.kr

Omics Technologies in Cancer Biomarker Discovery, edited by Xuewu Zhang.
©2011 Landes Bioscience.

They are present in tumor tissues or body fluids and encompass a wide variety of molecules, including transcription factors, cell surface receptors and secreted proteins. Effective tumor markers are in great demand since they have the potential to reduce cancer mortality rates by facilitating diagnosis of cancers at early stages and by helping to individualize treatment.

Glycosylation on the Protein

A recent report showed that the number of coding genes in human sequence (~21,000) is not dramatically different from the numbers reported for phylogenetically remote organism. There may be several hundred thousand human protein species after splice variants and essential post-translational modifications (PTMs) including glycosylation. Among more than 200 PTMs, it is estimated that protein glycosylation occurs over one-half of all mammalian proteins.[23] Protein glycosylation is also recognized as one of the most important PTM, which is involved in developmental biology, pathogen localization to host tissues, cell division, prion diseases, inflammation and tumor immunology.[24-29] Glycosylation has the potential to affect all stages of protein lifetimes, from folding and localization to the cell surface for their interactions with binding partners, to their degradation and their turnover. Covalent glycosylation occurs in a vast set of the proteins that are passing through endoplasmic reticulum in eukaryotic cells and then, go into other organelles of secretory pathway, including all the way to plasma membrane, as well as proteins secreted into the extracellular environments.[30] These diverse proteins may have a carbohydrate component that represent ranging from <1% to >80% of total weight. Sugars that commonly occur in glycoproteins include N-acetylglucosamine, mannose, glucose, galactose, N-acetylgalactosamine, sialic acid, fucose and xylose etc.

Mammalian glycoproteins contain three major types of oligosaccharides (glycans): N-linked glycans,[31] O-linked glycans[32] and glycosylphosphatidylinositol (GPI) lipid anchors.[33] The two major protein glycosylations are N-glycosylation and O-glycosylation. For N-linked oligosaccharides, a 14-sugar precursor core unit is first added to the asparagine in the polypeptide chain of the target protein, in the consensus tripeptide sequence Asn-X-Ser/Thr (Fig. 1). Among five different N-glycan linkages, this N-acetylglucosamine to asparagine (GlcNAcβ1-Asn) is the most common. The structure of this precursor is common to most eukaryotes and contains 2 N-acetylglucosamine, 9 mannose and 3 glucose molecules. A complex set of reactions attaches this branched chain to a carrier molecule called dolichol-phosphpate and then it is transferred to the appropriate point on the polypeptide chain as it is translocated into the ER lumen. It is not only the N-linked oligosaccharide chains on proteins that are altered as the proteins pass through the Golgi cisternae *en rout* from the ER to their final destinations. Proteins may contain multiple glycosylation sites that are modified with any of the three classes of N-linked glycans; high-mannose type, hybrid type and complex type (Fig. 2). Vertebrates have been found to possess a diverse compliment of complex and hybrid glycoproteins due to the broad variety of glycosidases and glycosyltransferases coded within the genome.[34] While these three classes of N-linked glycoproteins are also present in lower organisms, there is less diversity of structure than is found in vertebrate glycoproteins.

O-glycosylation is a common covalent modification on the serine and threonine residues of mammalian glycoproteins.[32] O-linked glycosylation is, thus, the modification of serine or threonine residues on the polypeptide by addition of a N-acetyl-galactosamine (GalNAc) residue. In mucins, O-glycans are covalently α-linked via an N-acetylgalactosamine (GalNAc) moiety to the -OH of serine or threonine by an O-glycosidic bond and the structures are named mucin O-glycans or O-GalNAc glycans.[35] The initiating event is the addition of the monosaccharide GalNAc (from UDP-GalNAc) to serine and threonine residues catalyzed by a polypeptide GalNAc transferase (GalNAcT). Also in contrast to N-glycosylation, a consensus sequence for GalNAc addition to polypeptides has not been found. In certain eukaryotes, disaccharide and branched tetra- and pentamannosyl chains are produced and it is estimated in those fungi that O-glycosylation is actually more common than the N-glycosylation of proteins. O-Linked glycoproteins are usually large proteins with a molecular mass of >200 kDa. While O-linkage does link primarily to peptide

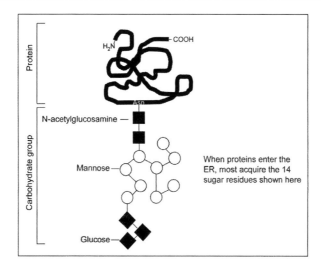

Figure 1. The structure of common N-linked glycosylation: Asparagine residue in protein is linked to N-acetylglucosamine and further core 14 sugar residues.

residues through a hydroxyl group, there is no consensus sequence required. O-linked glycans are commonly linear or biantennary and have comparatively less branching than N-glycans. It should be noted that there are also several types of nonmucin O-glycans including α-linked O-fucose, β-linked O-xylose, α-linked O-mannose, β-linked O-GlcNAc (*N*-acetylglucosamine), α- or β-linked O-galactose and α- or β-linked O-glucose glycans.

GPI anchored proteins are membrane-bound proteins and linked at their carboxy terminus through a phosphodiester linkage of phosphoethanolamine to a trimannosyl-nonacetylated glucosamine (Man3-GlcN) core.[33] The core may undergo various modifications during the secretion from cells. Their functionality ranges from enzymatic to antigenic and adhesive.

In general, glycosylation reactions are catalyzed by the actions of glycosyltransferases, sugar chains being added to various complex carbohydrates.[36] There are approximately over 180 glycosyltransferase genes of 300 known human genes (glycogenes), play a key role in the glycosylation process. Among them, N-Acetylglucosaminyltransferase family has been thought to have a close relationship with cancer metastasis.[37-39] Although the function of glycans on glycoproteins still remains elusive, there are emerging amounts of evidences accumulated in transgenic and knock-out mice regarding aberrant glycosylation which affects biological functions or leads to dysfunction.[37-39] Variation of oligosaccharides conjugated to proteins modulates the protein function by altering

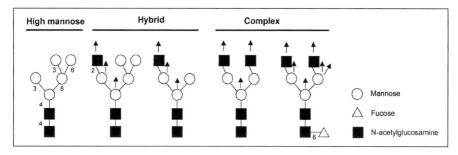

Figure 2. Major types of vertebrate N-glycans (Modified from a figure of "Carbohydrates and glycosylation" in Wormbook.org, edited by Patricia M. Berninsone.

protein folding, biological lifetime and the recognition of binding partners. Clinical relevance of glycosylation variation has been shown in inherited and nongenetic diseases, as a result of alterations in oligosaccharide structures.

Glycoproteome Analysis for Discovering Cancer Biomarkers

There is general agreement that early events in the evolution of neoplasia are involved in the alterations of oncogenes, tumor suppressor genes, mutator genes and apoptosis-related pathways. Subsequent tumor growth, invasion and metastasis is related with the survival of the fittest cells and it is therefore likely that the highly selective changes of glycosylation seen in tumor cells have the greatest functional consequences in these later stages.[38,40,41] Thus, aberrant glycosylation expressed in glycosphingolipids and glycoproteins in tumor cells also has been implicated as an essential mechanism in defining stage, direction and fate of tumor progression and certain glycan structures are well-known markers for tumor progression.[42] Especially, since disease-related glycoproteins are usually changed or released in blood or urine, these proteins has been highly considered as most potential useful biomarkers for the early diagnosis, monitoring and treatment of severe disease such as cancer.

Large amount of work in the area of biomarkers have made it clear that the efficacy of a given biomarker assay is determined by its sensitivity and specificity.[43-45] Both terms take on precise meanings in the development of biomarker tests for population-based screening or for clinic-based surveillance of high-risk population. Therefore, for a successful biomarker discovery, first of all, experimental design for searching targets should be established well.[46-48] However, the availability of high-quality, matched specimens has been limited and investigations are often compromised as a result since selected and validated potential biomarkers from the study in relatively small samples are not easy for researchers to associate with specific disease status due to variability and the factors in the followings; (1) Differences in methodology. (2) Lack of standardized sample collection and storage, variably affecting comparison groups. (3) Differences between cases and controls in terms of sex, age and physiological states (for example, fasting, weight gain or loss and hormonal status). (4) Differences in genetic make-up. (5) Changes in inflammation and acute-phase reactants. (6) Changes in metabolic states. (7) Other nonspecific changes, for example, cell death and tissue necrosis. (8) Changes reflecting underlying chronic disease, for example, those caused by smoking and chronic lung disease, in contrast to lung cancer-specific changes.[46]

Analysis of human body fluids including plasma/serum, urine, cerebrospinal fluid, saliva, bronchoalveolar lavage fluid, synovial fluid, nipple aspirate fluid, tear fluid and amniotic fluid has become one of the most promising approaches to the discovery of biomarkers for human diseases.[49-51] Various types of fluid and effusion offer a window onto the proteins from tumor tissue that may be released into extracellular fluids through secretion or cell and tissue breakdown. Based on this concept, recently, HUPO (the Human Proteome Organization formed in 2001) has launched a biological fluid project, namely the Plasma Proteome Project (PPP) for comprehensive analysis of plasma protein constituents in various states (disease, countries) of humans as well as normal of which in large cohorts of subjects although the dynamic range of known plasma proteins spans over ten orders of magnitude (Fig. 3).

Tumor-tissue-derived proteins in the circulation are probably present at the lower end of this range, particularly during the early stages of tumor development. Thus, enrichment of lower level proteins or of specific types of proteins has great advantage in the discovery of the cancer biomarkers in the human plasma. As described above, since the glycosylation patterns changes in the cancer development, glycoprotein enrichment can be a good way for the plasma proteome analysis. Usually, affinity chromatography methods using multiple lectins or multiple fractionations have been used to enrich the glycoproteins or peptides in plasma sample and these technical approaches are now providing good information for discovering biomarkers in cancers.[9,52,53]

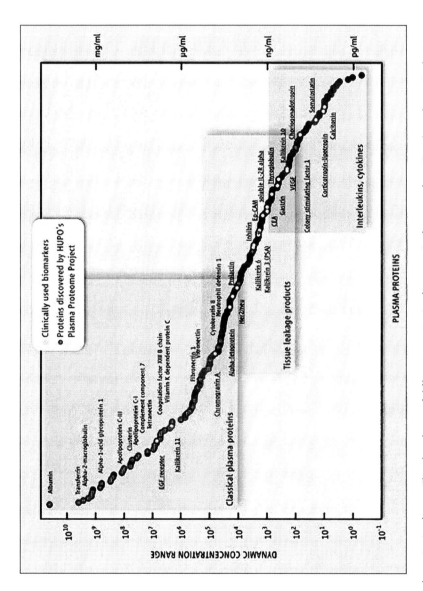

Figure 3. Protein dynamics in plasma. The abundance of different proteins in plasma varies by more than 10 orders of magnitude. Most commercially used biomarkers (yellow dots) are present in only minute quantities in blood, below the level at which most proteins are detected (red dots). (Adapted from Schiess R et al. Mol Oncol 2009; 3(1):33-44), ©2009 with permission from Elsevier. A color version of this image is available at www.landesbioscience.com/cuire.

Technologies in Glycoproteomics

In general, the MS-based technologies effectively used for protein identification are also applicable to the analysis of glycans.[54-56] However, glycoproteome analyses are inherently more difficult than simple protein/peptide identification for the following reasons; (1) Dynamic range of glycosylation modification is high. (2) Isolation of the modified peptide containing the glycosylation is required. (3) Glycosylations are frequently unstable. (4) Glycosylations are frequently transient in nature.

Indeed, most potential protein biomarkers are usually present at low levels and within very complex matrices although body fluids including serum, urine, or sputum have advantage for maximizing the usefulness and minimizing the cost for screening. To overcome these obstacles in general problem for discovering glycosylation-related biomarkers using proteomics techniques, advanced techniques have been developed recently.[57-59]

Purification Methods: Enrichment Techniques of Glycoproteins and Glycopeptides

Lectin Affinity Chromatography

Identification of glycosylated proteins or peptides requires separation, either physically or spectrometrically, from complex mixtures that contain both unglycosylated and heterogeneously glycosylated proteins or peptides. Historically, physical separation has been accomplished by affinity purification using lectins that are specific for specific glycan modifications.[60-62] Lectins are sugar-binding proteins which are highly specific for their sugar moieties. They may bind to a soluble carbohydrate or to a carbohydrate moiety which is a part of a glycoprotein or glycolipid. So, subjecting cell lysates or serum/plasma to lectin affinity chromatography enables enrichment of different classes of glycoproteins or glycopeptides and allows initial characterization of protein's glycan structure or the extent of glycosylation in the proteins under investigation. Classes of lectins for enrichment of various glycosylation types are listed as below: For the enrichemnt of *mannose glycoprotein,* ConA, GNA and LCH lectins are used, for *sialic glycoprotein,* WGA, SNA and MAL, for *O-Glycan glycoprotein,* AIL and PNA and for *total glycoprotein,* ConA and WGA.

Chemical Methods

Glycoproteins or glycopeptides also can be isolated on the basis of their chemical reactivity.[63,64] Several enzymatic or metabolic methods have been used to chemically tag proteins to enable their isolation. The most convergent approach is to append the carbohydrate(s) of interest to a synthetic peptide.

Immobilized Hydrazide

Hydrazides in organic chemistry are a class of organic compounds sharing a common functional group characterized by a nitrogen to nitrogen covalent bond with 4 substituents with at least one of them is acyl group.[65-67] The general structure for an hydrazide is (R1 = O)R2-N-N-R3R4. Hydrazide are useful for macromolecules at carbohydrate groups that have been oxidized to form aldehydes. The hydrazide group reacts with carbonyls (aldehydes and ketones), resulting in a hydrazone linkage.

General protocol for the analysis of glycoproteins by hydrizide immobilization is as follows:
1. Oxidation and covalent coupling: carbohydrate group of glycoproteins are oxidized into aldehydes by periodate and the oxidized glycoproteins are covalently coupled to hydrazide resin. No glycoproteins are removed by washing.
2. Proteolysis and Release of glycopeptides: glycoproteins on the beads are digested by trypsin and nonglycopeptides are removed by washing and retained glycopeptides on the resin are released by PNGase F.
3. Released glycopeptides are analyzed by LC-MS/MS and LC-FTICR.

Boronic Diester

Phenyl boronic acids are relatively stable organic metal compounds. They react with vicinal cis-diol group containing molecules whereby reversible covalent bonds to the boronic acid functionalities are formed. As a consequence, capturing reactions can be performed under mild alkaline conditions.[68,69] All glycostructures containing saccharides like mannose, galactose or glucose fulfill the requirements to form the pentamerous cyclic diester compound. Thereby, boronic acids react with glycoproteins containing N- and O-linked oligosaccharides.

Huisgen Cycloaddition
To date, the most popular reaction that has been adapted to fulfil these criteria is the 1, 3-dipolar cycloaddition, also known as "click chemistry", between azides and alkynes catalyzed by copper (I) salts.[70,71] The simplicity of this reaction and the ease of purification of the resulting products have opened new opportunities in enrichment of glycopeptides.

Antibody-Assisted Lectin Profiling (ALP)
A target protein is enriched from clinic samples (e.g., tissue extracts, cell supernatants or sera) by immuno-precipitation with a specific antibody recognizing a core protein moiety and the target glycoprotein is quantified by immuno-blotting using the same antibody.[72] Then, glycosylation difference is analyzed by means of antibody-overlay lectin microarray, an application technique of an emerging glycan profiling microarray.

Enrichemnt for GPI Proteins
Usually, m-Aminophenylboronic Acid Matrices and the immobilized lectins, Con A and Wheat Germ are used.

Glycoprotein Detection Methods
Initial detection of glycoproteins in vitro is routinely accomplished on SDS-PAGE gels and Western blots. Most of the commercially available modified glycoprotein detection tools are based on periodic acid-schiff (PAS) methods[73] and can detect sugar moieties of proteins in gels or on blot. There are many applied methods currently utilized in glycoprotein detection (for instance, fluorescent or biotin labeling).

Major components of colorimetric detection methods:
- **Oxidation component (periodic acid)**
- **Reduction component (sodium metabisulfite)**
- **Schiff's reagent, fuchsin-sulfite reagent**
- **Peroxidase**

Fluorescence method: In 2D-PAGE method, Glycoprotein Gel Stain Kit with SYPRO Ruby protein gel stain can provides a method for differentially staining glycosylated and nonglycosylated proteins in the same gel.[74]

Deglycoprotein Methods
Removal of glycans from glycoproteins is required to simplify the identification of glycoprotein or the analysis of the carbohydrates components.

Enzymatic Deglycosylation
For *N*-linked glycan: Use of the enzyme **PNGase F** is the most effective method of removing virtually all N-linked oligosaccharides from glycoproteins.[75,76] However, oligosaccharides containing a fucose a (1-3)-linked to the asparagine-linked N-acetylglucosamine, are resistant to PNGase F. N-Glycosidase A (PNGase A), isolated from almond meal, must be used in this situation. This enzyme, however, is ineffective when sialic acid is present on the N-linked oligosaccharide.

For *O*-linked glycan: There is no enzyme comparable to PNGase F for removing intact O-linked sugars. Monosaccharides must be sequentially hydrolyzed by a series of exoglycosidases until only the Gal-β(1-3)-GalNAc core remains. O-Glycosidase can then remove the core structure with no modification of the serine or threonine residue.[77]

For sequential deglycosylation: Endoglycosidases F1, F2 and F3 and Endoglycosidase H have been used for the analysis of individual monosaccharides from glycans and this can be useful for the elucidation of the structure and function of the glycan component.

Chemical Deglycosylation

Hydrazine hydrolysis has been found to be effective in the complete release of unreduced O- and N-linked oligosaccharides.[78] For O-linked glycans, alkaline β-elimination and Trifluoromethanesulfonic acid (TFMS) hydrolysis methods have also been used. For selective and sequential release of oligosaccharides, it can be accomplished by initial mild hydrazinolysis of the O-linked oligosaccharides at 60°C followed by N-linked oligosaccharides at 95°C.

Mass Spectrometry Analysis

Providing mass spectrometry (MS) as an accessible method for routine quantitative protein analysis was a quantum leap. Labeling techniques allow unbiased and multiplexed analysis of biological samples yielding a quantitative view on all the constituents of a complex and their changes. Biomarker search of a selected disease needs to be based on the quantity difference among samples. There are two different approaches in term of quantitative proteomics. One relies on the labeling methodology. The other is strictly based on mass spectra without labeling.

Bioinformatic Analysis of Glycoproteome

Bioinformatics is a very powerful tool in the field of glycoproteomics as well as genomics and proteomics. Some useful resources for prediction of glycosylation-sites or—structures on proteins are accessible on the web is as follows:

HGPI- http://www.hgpi.jp/menuA.html
Glycosuite-http://www.glycosuite.com
GlycoMod-http://expasy.ch/tool/glycomed
CarBank- http://www.boc.chem.uu.nl/sugabase/databases.html
NetNGlyc 1.0 Server-http://www.cbs.dtu.dk/services/NetNGlyc
Glyprot-http://www.glycosciences.de/modeling/glyprot/php/main.php

Validation of Glycoproteome

The glycoproteome identified by mass spectrometry needs to be validated by other technologies. Some important technologies are explained in the following.

Western Blot

Among identified potential biomarker, various iosforms of proteins can be validated by Western blot analysis. Since its identified partial part such as light- or heavy-chain of immunoglobulin like molecules or alterative splicing form of which can be detected by Western blot analysis. So this classical method using specific antibodies, is still most reliable technique to valid glycoproteins identified as potential biomarkers by proteomic analysis.

SRM (Single Reaction Monitoring)/MRM (Multiple Reaction Monitoring)

Western blotting may be tried-and-true, but SRM (or, as it is sometimes called, MRM, for multiple reaction monitoring) holds a significant advantage: it requires no antibodies and thus is amenable to studying proteins for which no antibodies exist. A mass spectrometric approach that monitors a molecule's abundance by its degradation into specific fragments employs tandem mass spectrometry to quantify selected proteins of interest, such as those previously identified in differential studies. Although, poor ionization problem of glycopeptides should be solved, one of the advantages of this approach is the multiplexing feature, allowing rapid characterization of many clinically relevant proteins simultaneously and in a high-throughput mode.[79]

ELISA

The most relied-on approach for validation remains the sandwich enzyme-linked immunosorbent assay (ELISA), which is highly specific because it uses a pair of antibodies against the targeted molecule including glycoproteins.[80,81]

ProteinChip (Protein Microarray or Microfluid Chip)

Arrays provide a variety of surface chemistries that allow researchers to perform high-throughput analysis. The surface chemistries of the arrays include a series of classic chromatographic chemistries or antibodies and specialized affinity capture surfaces.[82-84]

Conclusion

Most of the analytical approaches, including above those recent technologies, to discover cancer biomarkers, can be divided into major two fields. First, *Quantitative glycoproteomics* which identify specific glycoproteins expressed differentially in cancer milieu. Second, *Qualitative glycoproteomics* (*Glycomics or glycotyping*) which identify exchanged or modified sites of glycosylation on specific glycoproteins in cancer milieu.

It is plausible that highly advancing techniques for glycoproteome analysis will enable to discover biomarkers efficiently but, recently, it is proved that single biomarker is not able to provide the sensitivity and specificity required for most applications given the substantial heterogeneity among cancers. Therefore, groups or panels of specific glycoproteins are needed to obtain realistic sensitivity.[46] Moreover, the diverse modifications of glycan to specific glycoproteins in each cancer's states should also be established as a criterion.

References

1. FDA issues pharmacogenomics data submission guidance. Expert Rev Mol Diagn 2005; 5(3):275-277.
2. Sotiriou C, Pusztai L. Gene-expression signatures in breast cancer. N Engl J Med 2009; 360(8):790-800.
3. Gutman S, Kessler LG. The US Food and Drug Administration perspective on cancer biomarker development. Nat Rev Cancer 2006; 6(7):565-571.
4. Smith SC, Theodorescu D. Learning therapeutic lessons from metastasis suppressor proteins. Nat Rev Cancer 2009; 9(4):253-264.
5. Cho WC. Contribution of oncoproteomics to cancer biomarker discovery. Mol Cancer 2007; 6:25.
6. Koomen JM, Haura EB, Bepler G et al. Proteomic contributions to personalized cancer care. Mol Cell Proteomics 2008; 7(10):1780-1794.
7. Petricoin EF, Belluco C, Araujo RP et al. The blood peptidome: a higher dimension of information content for cancer biomarker discovery. Nat Rev Cancer 2006; 6(12):961-967.
8. Northen TR, Yanes O, Northen MT et al. Clathrate nanostructures for mass spectrometry. Nature 2007; 449(7165):1033-1036.
9. Heo SH, Lee SJ, Ryoo HM et al. Identification of putative serum glycoprotein biomarkers for human lung adenocarcinoma by multilectin affinity chromatography and LC-MS/MS. Proteomics 2007; 7(23):4292-4302.
10. Kuramitsu Y, Nakamura K. Proteomic analysis of cancer tissues: shedding light on carcinogenesis and possible biomarkers. Proteomics 2006; 6(20):5650-5661.
11. Cho JY, Sung HJ. Proteomic approaches in lung cancer biomarker development. Expert Rev Proteomics 2009; 6(1):27-42.
12. Bharti A, Ma PC, Salgia R. Biomarker discovery in lung cancer—promises and challenges of clinical proteomics. Mass Spectrom Rev 2007; 26(3):451-466.
13. Sung HJ, Cho JY. Biomarkers for the lung cancer diagnosis and their advances in proteomics. BMB Rep 2008; 41(9):615-625.
14. Kulasingam V, Diamandis EP. Strategies for discovering novel cancer biomarkers through utilization of emerging technologies. Nat Clin Pract Oncol 2008; 5(10):588-599.
15. Caron M, Choquet-Kastylevsky G, Joubert-Caron R. Cancer immunomics using autoantibody signatures for biomarker discovery. Mol Cell Proteomics 2007; 6(7):1115-1122.
16. Brusic V, Marina O, Wu CJ et al. Proteome informatics for cancer research: from molecules to clinic. Proteomics 2007; 7(6):976-991.
17. Hwang H, Zhang J, Chung KA et al. Glycoproteomics in neurodegenerative diseases. Mass Spectrom Rev 2009.
18. Packer NH, Harrison MJ. Glycobiology and proteomics: is mass spectrometry the Holy Grail? Electrophoresis 1998; 19(11):1872-1882.

19. Gevaert K, Impens F, Van Damme P et al. Applications of diagonal chromatography for proteome-wide characterization of protein modifications and activity-based analyses. FEBS J 2007; 274(24):6277-6289.
20. Banks RE, Dunn MJ, Hochstrasser DF et al. Proteomics: new perspectives, new biomedical opportunities. Lancet 2000; 356(9243):1749-1756.
21. Dwek MV, Ross HA, Leathem AJ. Proteome and glycosylation mapping identifies posttranslational modifications associated with aggressive breast cancer. Proteomics 2001; 1(6):756-762.
22. Morelle W, Canis K, Chirat F et al. The use of mass spectrometry for the proteomic analysis of glycosylation. Proteomics 2006; 6(14):3993-4015.
23. Apweiler R, Hermjakob H, Sharon N. On the frequency of protein glycosylation, as deduced from analysis of the SWISS-PROT database. Biochim Biophys Acta 1999; 1473(1):4-8.
24. Kornfeld S. Diseases of abnormal protein glycosylation: an emerging area. J Clin Invest 1998; 101(7):1293-1295.
25. Lowe JB. Glycosylation, immunity and autoimmunity. Cell 2001; 104(6):809-812.
26. Ohtsubo K, Marth JD, Glycosylation in cellular mechanisms of health and disease. Cell 2006; 126(5):855-867.
27. Rudd PM, Elliott T, Cresswell P et al. Glycosylation and the immune system. Science 2001; 291(5512):2370-2376.
28. Haltiwanger RS, Lowe JB. Role of glycosylation in development. Annu Rev Biochem 2004; 73:491-537.
29. Arnold JN, Wormald MR, Sim RB et al. The impact of glycosylation on the biological function and structure of human immunoglobulins. Annu Rev Immunol 2007; 25:21-50.
30. Fleischer B. Mechanism of glycosylation in the Golgi apparatus. J Histochem Cytochem 1983; 31(8):1033-1040.
31. Gu J, Sato Y, Kariya Y et al. A mutual regulation between cell-cell adhesion and N-glycosylation: implication of the bisecting GlcNAc for biological functions. J Proteome Res 2009; 8(2):431-435.
32. Comer FI, Hart GW. O-Glycosylation of nuclear and cytosolic proteins. Dynamic interplay between O-GlcNAc and O-phosphate. J Biol Chem 2000; 275(38):29179-29182.
33. Paulick MG, Bertozzi CR. The glycosylphosphatidylinositol anchor: a complex membrane-anchoring structure for proteins. Biochemistry 2008; 47(27):6991-7000.
34. Hinou H, Nishimura S. Mechanism-based probing, characterization and inhibitor design of glycosidases and glycosyltransferases. Curr Top Med Chem 2009; 9(1):106-116.
35. Brockhausen I. Mucin-type O-glycans in human colon and breast cancer: glycodynamics and functions. EMBO Rep 2006; 7(6):599-604.
36. Williams GJ, Thorson JS. Natural product glycosyltransferases: properties and applications. Adv Enzymol Relat Areas Mol Biol 2009; 76:55-119.
37. Pinho SS, Reis CA, Paredes J et al. The Role of N-acetylglucosaminyltransferase III and V in the Post-Transcriptional Modifications of E-cadherin. Hum Mol Genet 2009.
38. Zhao YY, Takahashi M, Gu JG et al. Functional roles of N-glycans in cell signaling and cell adhesion in cancer. Cancer Sci 2008; 99(7):1304-1310.
39. Kim YS, Hwang SY, Kang HY et al. Functional proteomics study reveals that N-Acetylglucosaminyl-transferase V reinforces the invasive/metastatic potential of colon cancer through aberrant glycosylation on tissue inhibitor of metalloproteinase-1. Mol Cell Proteomics 2008; 7(1):1-14.
40. Dube DH, Bertozzi CR. Glycans in cancer and inflammation—potential for therapeutics and diagnostics. Nat Rev Drug Discov 2005; 4(6):477-488.
41. Hakomori S. Glycosylation defining cancer malignancy: new wine in an old bottle. Proc Natl Acad Sci USA 2002; 99(16):10231-10233.
42. Zhao J, Patwa TH, Lubman DM et al. Protein biomarkers in cancer: natural glycoprotein microarray approaches. Curr Opin Mol Ther 2008; 10(6):602-610.
43. Gould Rothberg BE, Bracken MB, Rimm DL. Tissue biomarkers for prognosis in cutaneous melanoma: a systematic review and meta-analysis. J Natl Cancer Inst 2009; 101(7):452-474.
44. Vickers AJ, Jang K, Sargent D et al. Systematic review of statistical methods used in molecular marker studies in cancer. Cancer 2008; 112(8):1862-1868.
45. Simon R. Validation of pharmacogenomic biomarker classifiers for treatment selection. Cancer Biomark 2006; 2(3-4):89-96.
46. Hanash SM, Pitteri SJ, Faca VM. Mining the plasma proteome for cancer biomarkers. Nature 2008; 452(7187):571-579.
47. Suh KS, Remache YK, Patel JS et al. Informatics-guided procurement of patient samples for biomarker discovery projects in cancer research. Cell Tissue Bank 2009; 10(1):43-48.
48. Shah S. Biomarkers—select biosciences' third European summit. IDrugs 2008; 11(12):876-879.
49. Good DM, Thongboonkerd V, Novak J et al. Body fluid proteomics for biomarker discovery: lessons from the past hold the key to success in the future. J Proteome Res 2007; 6(12):4549-4555.

50. Hu S, Loo JA, Wong DT. Human body fluid proteome analysis. Proteomics 2006; 6(23):6326-6353.
51. Bergquist J, Palmblad M, Wetterhall M et al. Peptide mapping of proteins in human body fluids using electrospray ionization Fourier transform ion cyclotron resonance mass spectrometry. Mass Spectrom Rev 2002; 21(1):2-15.
52. Jung K, Cho W, Regnier FE. Glycoproteomics of plasma based on narrow selectivity lectin affinity chromatography. J Proteome Res 2009; 8(2):643-650.
53. Orazine CI, Hincapie M, Hancock WS et al. A proteomic analysis of the plasma glycoproteins of a MCF-7 mouse xenograft: a model system for the detection of tumor markers. J Proteome Res 2008; 7(4):1542-1554.
54. Seeberger PH, Werz DB. Synthesis and medical applications of oligosaccharides. Nature 2007; 446(7139):1046-1051.
55. Raman R, Raguram S, Venkataraman G et al. Glycomics: an integrated systems approach to structure-function relationships of glycans. Nat Methods 2005; 2(11):817-824.
56. Drickamer K, Taylor ME. Glycan arrays for functional glycomics. Genome Biol 2002; 3(12):REVIEWS 1034.
57. Block TM, Comunale MA, Lowman M et al. Use of targeted glycoproteomics to identify serum glycoproteins that correlate with liver cancer in woodchucks and humans. Proc Natl Acad Sci USA 2005; 102(3):779-784.
58. Lim KT, Miyazaki K, Kimura N et al. Clinical application of functional glycoproteomics—dissection of glycotopes carried by soluble CD44 variants in sera of patients with cancers. Proteomics 2008; 8(16):3263-3273.
59. Meany DL, Zhang Z, Sokoll LJ et al. Glycoproteomics for prostate cancer detection: changes in serum PSA glycosylation patterns. J Proteome Res 2009; 8(2):613-619.
60. Fang X, Zhang WW. Affinity separation and enrichment methods in proteomic analysis. J Proteomics 2008; 71(3):284-303.
61. Dayarathna MK, Hancock WS, Hincapie M. A two step fractionation approach for plasma proteomics using immunodepletion of abundant proteins and multi-lectin affinity chromatography: application to the analysis of obesity, diabetes and hypertension diseases. J Sep Sci 2008; 31(6-7):1156-1166.
62. Taniguchi N, Ekuni A, Ko JH et al. A glycomic approach to the identification and characterization of glycoprotein function in cells transfected with glycosyltransferase genes. Proteomics 2001; 1(2):239-247.
63. Zhang H. Glycoproteomics using chemical immobilization. Curr Protoc Protein Sci. 2007 May;Chapter 24:Unit 24.3.
64. Sun B, Ranish JA, Utleg AG et al. Shotgun glycopeptide capture approach coupled with mass spectrometry for comprehensive glycoproteomics. Mol Cell Proteomics 2007; 6(1):141-149.
65. Lewandrowski U, Sickmann A. N-glycosylation site analysis of human platelet proteins by hydrazide affinity capturing and LC-MS/MS. Methods Mol Biol 2009; 534:225-238.
66. Cao J, Shen C, Wang H et al. Identification of N-glycosylation sites on secreted proteins of human hepatocellular carcinoma cells with a complementary proteomics approach. J Proteome Res 2009; 8(2):662-672.
67. McDonald CA, Yang JY, Marathe V et al. Combining results from lectin affinity chromatography and glycocapture approaches substantially improves the coverage of the glycoproteome. Mol Cell Proteomics 2009; 8(2):287-301.
68. Zhang Q, Tang N, Brock JW et al. Enrichment and analysis of nonenzymatically glycated peptides: boronate affinity chromatography coupled with electron-transfer dissociation mass spectrometry. J Proteome Res 2007; 6(6):2323-2330.
69. Gontarev S, Shmanai V, Frey SK et al. Application of phenylboronic acid modified hydrogel affinity chips for high-throughput mass spectrometric analysis of glycated proteins. Rapid Commun Mass Spectrom 2007; 21(1):1-6.
70. Clark PM, Dweck JF, Mason DE et al. Direct in-gel fluorescence detection and cellular imaging of O-GlcNAc-modified proteins. J Am Chem Soc 2008; 130(35):11576-11577.
71. Gurcel C, Vercoutter-Edouart AS, Fonbonne C et al. Identification of new O-GlcNAc modified proteins using a click-chemistry-based tagging. Anal Bioanal Chem 2008; 390(8):2089-2097.
72. Kuno A, Kato Y, Matsuda A et al. Focused differential glycan analysis with the platform antibody-assisted lectin profiling for glycan-related biomarker verification. Mol Cell Proteomics 2009; 8(1):99-108.
73. Devine PL, Warren JA, Layton GT. Glycoprotein detection using the periodic acid/Schiff reagent: a microassay using microtitration plates. Biotechniques 1990; 8(4):354-356.
74. Patton WF. Fluorescence detection of glycoproteins in gels and on electroblots. Curr Protoc Cell Biol 2002; Chapter 6: Unit 6.8.
75. Morelle W, Faid V, Chirat F et al. Analysis of N- and O-linked glycans from glycoproteins using maldi-tof mass spectrometry. Methods Mol Biol 2009; 534:5-21.

76. Suzuki T, Park H, Lennarz WJ. Cytoplasmic peptide:N-glycanase (PNGase) in eukaryotic cells: occurrence, primary structure and potential functions. FASEB J 2002; 16(7):635-641.
77. Kondoh G, Watanabe H, Tashima Y et al. Testicular Angiotensin-converting enzyme with different glycan modification: characterization on glycosylphosphatidylinositol-anchored protein releasing and dipeptidase activities. J Biochem 2009; 145(1):115-121.
78. Babizhayev MA, Guiotto A, Kasus-Jacobi A. N-Acetylcarnosine and histidyl-hydrazide are potent agents for multitargeted ophthalmic therapy of senile cataracts and diabetic ocular complications. J Drug Target 2009; 17(1):36-63.
79. Mollah S, Wertz IE, Phung Q et al. Targeted mass spectrometric strategy for global mapping of ubiquitination on proteins. Rapid Commun Mass Spectrom 2007; 21(20):3357-3364.
80. Koyama H, Yamamoto H, Nishizawa Y. RAGE and soluble RAGE: potential therapeutic targets for cardiovascular diseases. Mol Med 2007; 13(11-12):625-635.
81. Onorato JM, Thorpe SR, Baynes JW. Immunohistochemical and ELISA assays for biomarkers of oxidative stress in aging and disease. Ann N Y Acad Sci 1998; 854:277-290.
82. Chu CS, Ninonuevo MR, Clowers BH et al. Profile of native N-linked glycan structures from human serum using high performance liquid chromatography on a microfluidic chip and time-of-flight mass spectrometry. Proteomics 2009; 9(7):1939-1951.
83. Alley WR Jr, Mechref Y, Novotny MV. Use of activated graphitized carbon chips for liquid chromatography/mass spectrometric and tandem mass spectrometric analysis of tryptic glycopeptides. Rapid Commun Mass Spectrom 2009; 23(4):495-505.
84. Dayal B, Ertel NH. ProteinChip technology: a new and facile method for the identification and measurement of high-density lipoproteins apoA-I and apoA-II and their glycosylated products in patients with diabetes and cardiovascular disease. J Proteome Res 2002; 1(4):375-380.

CHAPTER 9

Lipidomics in Cancer Biomarker Discovery

Xiaoqiong Ma and Jun Yang*

Abstract

Recent development has revealed that besides a role as important structural components, lipids also have important physiological functions involved in many cellular signal transduction pathways such as cell growth, differentiation and cell death. The abnormal metabolism of lipids is closely related to many human diseases. In particular the incidence of malignant tumors is increased in overweight and obes persons caused by the so-called metabolism syndrome. Lipids, including fatty acids, triglycerides, cholesterol, phosphoglycerides, sphingomyelins, glycosphingolipid, ceramide and sphingosine phosphate, were found to be associated with malignant tumors. The total lipid composition of a cell or organ is defined as the "lipidome" and the study of lipids and their interacting partners is termed "lipidomics". With the rapid development of lipidomics techniques, more and more attention has been paid to lipids because of their important roles in tumor diagnosis and therapy.

Introduction

Lipids are the metabolites of organism as well as the absolute necessary component materials of organism. They provide the energy for metabolism and transport biomacromolecules including proteins, fatty acids and vitamins in lipid-enclosed vesicles within the cell. Lipids are most enriched in cell membrane and can form "lipid raft" to provide an interaction platform for membrane proteins. Through various combinations of fatty acids with backbone structures, an enormous number of structurally distinct lipid molecular species could be generated theoretically, even though the nonrandom distribution of chemical structure may render a much lower number in reality.[1] Just like the definition for "genome" or "proteome", the total lipid composition of a cell or organ is defined as the "lipidome".

The abnormal metabolism of lipids has been shown to be closely related to many human diseases. For example, the oxidation of membrane lipids is a common process in many chronic inflammation diseases;[2] and the elevated levels of isoprostanes could be observed in carbon tetrachloride hepatotoxicity, diabetes mellitus, Alzheimer's disease, tobacco abuse, renal failure, heart failure and atherosclerosis, thus indicating that the metabolites of lipids could be used as biomarker for diseases.[3] However, for a long time, the study on lipids had been largely ignored. The main reason was the lack of recognition for the functional/physiological importance of lipids. Moreover, the lack of suitable tools for analyzing the complex lipids also posed as a huge obstacle for lipid study. In lipids analysis, the traditional methods, including isotope labeling, thin layer chromatography (TLC) and high performance liquid chromatography (HPLC), all have certain limitations such as low sensitivity and long experimentation time. The development of mass spectrometry (MS)

*Corresponding Author: Jun Yang—Department of Nutrition and Toxicology, Zhejiang University School of Public Health, Hangzhou, Zhejiang, 310058 China. Email: gastate@zju.edu.cn

Omics Technologies in Cancer Biomarker Discovery, edited by Xuewu Zhang.
©2011 Landes Bioscience.

technology, especially the appearance of soft ionization techniques (such as matrix-assisted laser desorption ionization (MALDI) and electro-spray ionization (ESI)) greatly improved our ability to analyze lipids. These MS methods have made the high-throughput analysis of lipid possible and in a faster and more accurate way. Thus, lipid study has reached at the systems level. The word "lipidomics" was first coined in 2003 and it was defined as the emerging field of systems-level analysis of lipids and factors that interact with lipids.[4,5]

Abnormal lipid metabolism could lead to overweight and obesity. Epidemiological evidence has shown that excess body weight is a risk factor for several cancer types, including cancer of the colon, breast, endometrium, kidney, esophagus, as well as possibly additional sites.[6] In particular, cancer is more frequently observed in obesity persons with Metabolism syndrome.[7] Since lipid metabolism is changed in cancer patients, using lipidomic techniques to compare the lipid profiles of health persons and cancer patients could lead to the identification of those lipid species with altered expression. These specific lipid molecules then might be used as potential biomarkers for cancer, which could help in the diagnosis and treatment of cancer, as well as providing prognostic information and the characterization of the pathologic features of cancers.

As mentioned above, the study on lipids is far behind those for genes and proteins. Even the classification of lipids remained to be a problem as recent as 2005. At that year, a new nomenclature system for lipids has been proposed, in which lipids were divided into eight major categories: (1) fatty acyls, (2) glycerolipids, (3) glycerophospholipids, (4) sphingolipids, (5) sterol lipids, (6) prenol lipids, (7) saccharolipids and (8) polyketides.[8] In this chapter, we will focus on some lipids that have been shown to be associated with cancer. These include fatty acid (belong to fatty acyls), triglyceride (glycerolipids), cholesterol (sterol lipids), phosphoglyceride (glycerophospholipids), sphingomyelin (sphingolipids), glycosphingolipid (sphingolipids and saccharolipids), ceramide (sphingolipids) and sphingosine phosphate (sphingolipids). Their structures are shown in Figure 1.

Lipids and Cancer

Fatty Acids

Fatty acids have a long hydrocarbon chain and a carbonyl at an end, the hydrocarbon chains are mainly linear, few are ramose or ringed. Some of the hydrocarbon chains are saturated and some are unsaturated. Fatty acids are also called eicosanoids, which include free fatty acids, fatty acid amides, prostanoids, hydroxyl-and hydroperoxy-eicosaenoic acids, leukotrienes and exoxyeicosatrienoic acids. Polyunsaturated free fatty acids participate in normal functioning of the cell, particularly involved in intracellular signaling. The level of fatty acids has been reported to change in the process of cancer. For example, it has been shown that in hepatocellular carcinoma (HCC) patients, the plasma free fatty acids levels were slightly or significantly decreased.[9] By comparing human meningiomas and normal leptomeninges, it was found that unsaturated fatty acids increased, while saturated fatty acids decreased in meningiomas.[10] In colorectal cancer, it was shown that the malignant transformation was accompanied by a decrease in the amount of linoleic acid (LA) and alpha-linolenic acid (ALA), while arachidonic and oleic acids increased.[11]

Triglycerides

Triglycerides are compounds of glycerol and fatty acids in which three hydroxyl of glycerol were esterified by fatty acids. Triglycerides levels in blood plasma are generally used to determine the health state of a person. Accumulating data have shown that triglycerides levels in plasma are associated with various types of cancer. In the study of plasma lipid profiles of gynecologic cancer patients, it was found that in breast cancer patients there is a moderate increase in the plasma levels of triglycerides (18%); in ovarian cancer patients, there is a high decrease in the plasma levels of triglycerides (31%); in gynecologic cancers other than breast and ovarian cancer, there is a moderate decrease in plasma levels of triglycerides (25%).[12] In addition, plasma levels of triglycerides could discriminate between patients with benign breast disease and breast cancer patients, as higher triglycerides levels were associated with increased breast cancer risk.[13] Plasma lipid profiles could also change in HCC and in the majority of reports about HCC, plasma levels

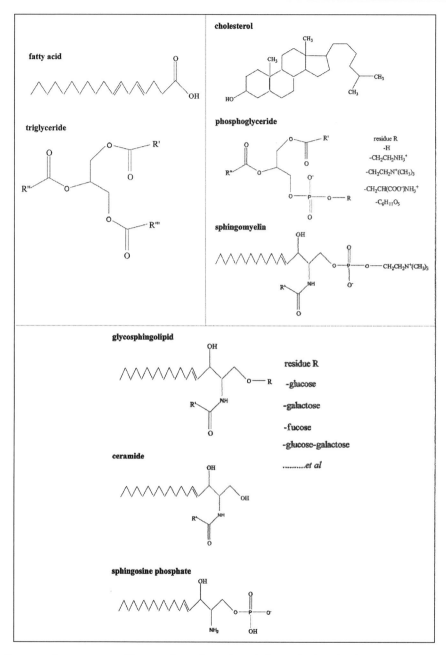

Figure 1. The structures of fatty acids, triglycerides, cholesterol, phosphoglycerides, sphingo-myelins, glycosphingolipid, ceramide and sphingosine phosphate.

of triglycerides were slightly to significantly decreased, while in certain cases plasma levels of triglycerides might be increased.[9] The plasma triglycerides levels were also significantly lower in untreated head and neck cancer patients compared to patients with oral precancerous conditions (OPC) and controls.[14] On the contrary, in colorectal carcinoma the serum triglyceride levels tend to increase. By examining the relationship between serum triglycerides and colorectal carcinoma in situ, it was found that serum triglyceride levels were significantly and positively associated with colorectal carcinoma in situ risk.[15]

Cholesterol

Cholesterol levels in blood plasma are also used to determine the health state of a person and there also exists a link between plasma cholesterol levels and cancer risk. Just as triglycerides in gynecologic cancer, in breast cancer patients there is moderate increase in the plasma levels of cholesterol (21%) and a high increase in LDL-cholesterol (43%), while there is a moderate decrease in HDL-cholesterol levels (30%) when compared to normal subjects.[12] In benign breast disease, total cholesterol was also significantly lower compared with controls.[13] Similarly, in ovarian cancer patients, there is a high decrease in the plasma levels of cholesterol when compared with normal subjects.[12,16] In gynecologic cancers other than breast and ovarian cancer, there is a moderate decrease in plasma levels of the cholesterol (21%) and HDL-cholesterol levels (27%), while a nonsignificant decrease in LDL-cholesterol (6.2%) when compared with normal subjects.[12] Plasma cholesterol levels also decreased in HCC and decreased serum levels of cholesterol may indicate a poor prognosis.[9] In untreated head and neck cancer patients, a significant decrease in plasma total cholesterol and HDL-cholesterol was observed.[14] In nondiabetic prostate carcinoma cases, LDL-cholesterol was significantly increased and HDL-cholesterol was significantly decreased compared to controls.[17] Finally, there was also a significant positive association between serum total cholesterol levels and the risk of colorectal carcinoma.[15]

Phosphoglycerides

Phosphoglycerides are the main components of biological membrane. Phosphoglycerides have two hydrocarbon chains in the *sn*-1 and *sn*-2 positions of the glycerol backbone and one phosphoric acid or one phosphoric acid esterified group in the *sn*-3 positions. There are many types of phosphoglycerides, including phosphatidic acid (PA), phosphatidylcholine (PC), phosphatidylethanolamine (PE), phosphatidylglycerol (PG), phosphatidylserine (PS), phosphatidylinositol (PI), cardiolipin (CL), lysophospholipids, plasmalogens and other ether-linked phospholipids.

Many studies have shown that lysophosphatidic acid (LPA) could be used as a potential biomarker for cancer. For example, in ovarian cancer, significantly higher total LPA levels were determined in the serum of patients with different types of tumors (benign and malignant).[18-21] In other gynecologic cancers, patients also showed higher LPA levels compared with healthy controls.[22] In contrast, lysophosphatidylcholines (LPC) levels are usually decreased in the blood plasma of cancer patients. For instance, lower concentration of LPC was observed in the blood plasma of renal cell carcinoma patients;[23] and serum LPC levels were significantly decreased in acute leukemia, malignant lymphomas, early stages of digestive tract tumors and multiple myeloma,[24,25] as well as in colorectal cancer.[26] Similarly, PC, PE, PS and PI levels in serum also decrease in cancer patient. In acute leukemia, PC, PE, PI and some other phospholipids levels were significantly lower than in reference group.[27] Based on such information, de Castro et al examined the levels of phosphoglycerides in erythrocytes and platelets of advanced nonsmall cell lung cancer patients, in an effort to evaluate the possibility of using phosphoglycerides as potential cancer biomarker; and the results showed that using certain combination of these phosphoglycerides have similar/higher diagnostic yields than the commonly used markers sialic acid or cytokeratins.[28]

Phosphoglycerides also change in cancer tissues. In esophageal cancer tissues, PS, PI and PC levels differed significantly from normal esophagus.[29] In estrogen receptor-positive breast tumor, the levels of alkylacylphosphatidylcholine (AAPC) were elevated whereas the levels of PC were diminished.[30] However, it has been reported that PC accumulated in human breast cancer cell-lines compared to normal human mammary epithelial cells,[31] thus further studies are needed to clarify this discrepancy. In colon cancer, increased choline kinase activity and PC levels were observed in colon cancer and adenoma tissue.[32] Merchant et al also reported that LPC and PC plasmalogen were significantly elevated in malignant colon cancer tissues along with significantly decreased levels of PE plasmalogen.[33] Similarly, malignant neoplasm cells with metastases are characterized by a higher PC/PE ratio than malignant neoplasm cells without metastases.[34] For neural tumor, changes in phosphoglycerides have also been reported. In human meningiomas, PI and PC levels are increased, while alkenyl-acyl PE levels are decreased.[10] And in human gliomas, there is a gradual increase of PC/PE ratio with the malignancy degree.[35]

Sphingomyelins

Sphingomyelins are the ramification products of sphingosine. In sphingomyelins, the amido of sphingosine is connected to a fatty acid by amido bond and the hydroxide of sphingosine is connected to a phosphorylcholine by ester bond. Sphingomyelins are mainly located at cell membrane and other lipid abundant tissues. Sphingomyelins have important functions in maintaining cell membrane structure, as well as transferring signals through membrane.

Alteration in sphingomyelins is also observed in cancer, either in blood plasma or cancer tissues. In acute leukemia patient plasma, the concentrations of sphingomyelins were lower than those of the reference group;[27] while in breast tumor patients, blood plasma sphingomyelin levels were increased.[36] In cancer tissues, significant elevations of sphingomyelins were noted malignant breast tumors and colon tumors.[30,37] Elevation of sphingomyelin level was also noted in all analyzed patterns in pulmonary tissue from lung cancer patients.[38] However, in human meningiomas, a decrease of sphingomyelin was detected.[10] Interestingly, animal studies have shown that adding sphingomyelins to diets have significant anticancer function,[39-41] indicating the potential application of sphingomyelins in cancer prevention and therapy.

Glycosphingolipids

Glycosphingolipids are also the ramification products of sphingosine. In glycosphingolipids, the amido of sphingosine is connected to a fatty acid by amido bond and the hydroxide of sphingosine is connected to monosaccharide or polysaccharide by glycosidic bond. There are neutral and acid types of glycosphingolipids, in which cerebrosides and gangliosides represent each type. Sulfatides belong to cerebrosides, its glycosyl carbon atom C3 is esterified by sulfuric acid. Gangliosides are also called sialoglycosphingolipid, as they have a sialic acid in their glycosyls. Glycosphingolipids have many biological functions and the changes of glycosphingolipids, especially gangliosides, in cancer have been studied more thoroughly than other lipid species.

As early as 1976, Watanabe et al had observed that the concentration of cerebroside alpha-L-fucopyranosylceramide increased in colon tumor tissue.[42] And gangliosides have been regarded as antigens of melanoma in the 1980s,[43] as ganglioside GM2, GD2 and GD3 often became major components of cultured melanoma cells.[44,45] However, the ganglioside profiles in melanoma tissues differed among different sites of tumor, pigmentation and histopathologic types. They were also different from cultured melanoma cell lines. Only GM3 was positively correlated with a good prognosis in both biopsy samples and cultured melanoma cells.[46]

In renal cell carcinoma, the inverse relationship of expression between GM3 and globo-series ganglioside is reflected on the degree of malignancy of renal cell carcinoma, the level of higher gangliosides is correlated with the degree of metastatic potential.[47-49] Later study has shown that renal cell carcinoma patients with tumors positive for GalNAc disialosyl lactotetraosylceramide (GalNAcDSLc4) are at higher risk of metastasis at the time of diagnosis and during follow-up.[50]

For ovarian cancer, the level of globotriaosylceramide Gb3 was low or even undetectable in normal ovarian tissue; while it was significantly increased in both benign and malignant tumors, with the highest Gb3 content observed for secondary ovarian metastases tumors and tumors refractory to chemotherapy.[51] GM3 and GD3 have been reported as the main tumor gangliosides in different types of human ovary benign and malignant epithelial tumors. However, their concentrations decrease in malignant ovary tumors, especially the amount of ganglioside GD3.[52]

In prostatic cancer tissue, the amounts of lactosyl and globoside series glycolipids were found to be generally reduced.[53] Ohyama et al have suggested that Gb3 may be a useful histological marker for testicular germ cell tumors.[54] Furthermore, marked accumulation of globotriaosyl ceramide was observed in seminoma but it was present in a small amount in testicular malignant lymphoma, thus it could be used to distinguish seminoma and malignant lymphoma.[55] On the other hand, galactosylgloboside Gb5 was identified in all nonmetastatic seminomas and lacked in all metastatic seminomas, implying the application of Gb5 to distinguish metastases tumor.[56]

Gangliosides are widely existed in neural systems and many studies focusing on gangliosides in neural and brain tumors have been published. In human brain tumors, it has been reported that patterns of gangliosides and neutral glycolipid could be of considerable value in refining the classification, diagnosis and prognosis of human brain tumors.[57-60] For example, 1b gangliosides (GD1b, GT1b and GQ1b) and 6'-LM1 (NeuAc alpha 2—>6Gal beta 1—>4Glc-NAc beta 1—>3Gal beta 1—>4Glc beta 1—>1Cer) may be used as prognostic indicators for primary brain tumor.[61] GD1b may also be used as a diagnostic and prognostic marker for some primary human brain tumors, especially astrocytomas.[62] Combining GM1 with galactosytransferase and paragloboside could be used to distinguish astrocytic and oligodendroglial tumors.[63]

Similarly, increased sialyllactotetraosylceramide has been suggested as a ganglioside marker for human malignant gliomas.[64] A gradual accumulation of lattosylceramide and GD3 could also constitute a valid marker of the malignancy grade of human gliomas.[35] On the other hand, galactosylceramide and sulfatide were absent in malignant gliomas.[65] In different types of ependymal tumors, glycolipid compositions are also different from each other[66] and high proportions of globoside and low ratios of GD1a:GD1b may be the markers for pilocytic astrocytomas and pleomorphic xanthoastrocytomas.[67]

Besides as potential cancer markers, the possible application of glycosphingolipids in cancer treatment is also of interest. Using various colorectal carcinoma and gastric cancer cell lines, Ono et al have found that GM3 interacts with CD9 molecule and that CD9 and GM3 cooperatively down-regulate tumor cell motility.[68]

Ceramide

Ceramide is another group of the ramification products of sphingosine and is the base structure of sphingomyelins. In ceramides, the amido of sphingosine is connected to a fatty acid by amido bond. Ceramide is at the center of sphingolipids metabolism and has important functions in many cellular events, including cell growth, senescence, meiotic maturation and cell death.[69] There have been many research reports focusing on the molecular functions and the metabolic pathway of ceramide. The most studied function of ceramide is its ability to induce apoptosis. Ceramide generated in the cell membrane is central for the induction of apoptosis by death receptors and many stress stimuli, such as UV, gamma-irradiation, or pathogen infection;[70] and ceramide could induce apoptosis in almost any type of cells,[71] including many cancer cells such as breast cancer cell (MCF-7),[72] colorectal cancer cell (L0V0, HT-29, HCT-116),[73-75] prostate cancer cell (LNCaP)[76] and Glioblastoma multiforme (GBM) cell.[77]

Changes in the ceramide metabolism are also closely related to many human diseases such as neurological disorders, cancer, infectious diseases and Wilson's disease.[70] For example, ceramide levels were decreased in malignant progression tumors, such as human ovary epithelial tumors[52] and human glial tumors.[78] Therefore, modulation of ceramide metabolism has been considered a target for cancer therapy.[79,80] Indeed, several anticancer agents, including the cytotoxic retinoid fenretinide (4-HPR), can act, at least partially, by increasing tumor cell ceramide via de novo synthesis.[79] A number of enzymes involved in ceramide metabolism are beginning to be recognized as potential targets for cancer therapy.[81] Controlling the balance between ceramide and sphingosine-1-phosphate (see below) also represents a promising rational approach to effective cancer therapy.[82,83]

Sphingosine Phosphate

Sphingosine-1-phosphate (S1P) is a polar lysophospholipid metabolite that is stored in platelets and released upon their activation. Although most members of the sphingolipid family execute negative effects on cell growth, S1P is able to induce cell proliferation and protect cells from undergoing apoptosis.[84-87] S1P can prevent ceramide-induced apoptosis in several cell types and doing so by inhibiting the activation of caspases.[88,89] Treatment of murine fibroblast C3H10T1/2 cells with ceramide does not induce apoptosis, which is attributed to the conversion of the pro-apoptotic ceramide to the anti-apoptotic metabolite S1P.[90] S1P can also regulate cytoskeletal remodeling, cellular motility and metastatic invasiveness as well.[91-93] As S1P is generated from ceramide by the consecutive actions of ceramidase and sphingosine kinase, a delicate balance between ceramide and S1P will determine whether cells undergo apoptosis or proliferate. Thus, this balance provides a novel target for cancer therapy.[82,83,94]

Lipidomics Methodology

There are several experimental approaches could be used in lipid research, such as chromatography, mass spectrometry, nuclear magnetic resonance (NMR) and biochemistry methods. NMR could be used to analysis lipid directly in a nondestructive manner, however, the sensitivity of NMR is low and could only determined very abundant and dominate lipids (such as cholesterol and PC). Biochemistry methods could analysis many lipids either in pure form or in mixtures, but biochemistry methods usually are experimentally and technically challenging and some limitations such as sometimes special lipid antibodies are required, which could limit its application.[1] Here we mainly discuss chromatography and mass spectrometry methods in lipidomics.

Lipid Extraction

The first step in lipidomic analysis is the extraction of lipid. Almost all lipid extraction methods use organic solvents since lipid has a high solubility in organic solution and lipid separation is created between immiscible solvents for lipids partitioning into the hydrophobic phase. Most of the current extraction methods are based on the methods established in the 1950s, in which chloroform, methanol and water are used.[95,96] In their methods, the lower phases are pooled to obtain the total lipid extract. Many modified extraction methods have since been established, such as using varying proportion of chloroform and methanol,[97] using isopropanol/hexane to extract more prostaglandin,[98] using icing LiCl solution to homogenize tissues before adding chloroform and methanol[99] and for extracting a specific class of lipid such as sphingolipids.[100]

Lipid Separation

Chromatography is the main method to separate lipid. Thin-layer chromatography (TLC) was used in lipid separation in the 1960s.[101] Solid-phase extraction (SPE) chromatography is very useful in the separation of crude lipid mixtures into different lipid classes. Octadecylsilyl SPE columns have been used to isolate arachidonic acid metabolites from biological samples[102,103] and aminopropyl SPE columns have been used to isolate phospholipids, fatty acids, cholesteryl esters, monoglycerides, diglycerides, triglycerides and cholesterol froma crude lipid mixture.[104] High performance liquid chromatography (HPLC) is the most common separation method used for lipids.[105-109] Fatty acids are nearly always separated on a reverse-phase ODS (C18) column using a methanol or acetonitrile based gradient solvent system in water; columns containing a chiral stationary phase can effectively separate enantiomers of lipids such as hydroxyeicosatetraenoic acids; and separation of phospholipids can be achieved by either normal-phase HPLC or reverse-phase HPLC.[3]

Lipid Mass Spectrum Analysis

Mass spectrometry is a technique that performs the molecular identification through determining the ratio of mass to charge (m/z) of a molecular. Gas chromatography (GC) tandem mass spectrometry has been used in lipid analysis for several decades. The prerequisite condition of GC analysis is the volatility of lipid under GC working temperature. Because many lipids are nonvolatile in nature state, volatile derivatives of lipids were obtained by chemical derivation for analysis.[110-112] The development of soft ionization techniques, including matrix-assisted laser desorption ionization (MALDI) and electrospray ionization (ESI), have revolutionized the mass spectrometry technique for lipid analysis and the volatility of lipid is no longer an issue. Lipid can be directly ionized and analyzed by mass spectrometer, which greatly improved the speed and accuracy and in particular, the high-throughput lipid analysis makes lipidomic research possible.

MALDI-time of flight mass spectrometry (MALDI-TOF MS) is a laser-based soft-ionization method often used for protein analysis; however, it has also been successfully used for lipids. The lipid sample is mixed with a matrix, usually 2,5-dihydroxybenzoic acid and applied to a sample plate and dried.[113] Then a 337 nm nitrogen laser is fired at the dried spot, the matrix absorbs the energy, which is then transferred to the lipid, resulting in the ionization of lipid molecule. Both positive-ion and negative-ion mode could be used to obtain positive molecular ion peak $(M+H)^+$ or $(M+Na)^+$, as well as negative molecular ion peak $(M-H)^-$. The molecular ion peaks are then used to identify lipid species. The sample plate can hold dozens of samples, with nano-gram sample is adequate for analysis and every spot can be analyzed within 1 minute. However, to accurately determine the structure of lipid molecule, sometimes tandem MS is needed. Still, as a fast and comparatively easier method, MALDI-TOF is widely used by many researchers for lipid analysis. For example, Fujiwaki et al has reported lipid profile changes in Gaucher, Niemann-Pick, Fabry and sphingolipidosis diseases.[100,114-117] similarly, chemical-induced changes of sphingolipid profile in cell or animal model have been studied, which showed dramatic effects of various chemicals on sphingolipid metabolism.[118-120] Direct determination of lipid distribution in brain tissues has also been conducted using MALDI-TOF or MALDI-TOF-TOF.[121-124]

ESI can analyze nonvolatile molecules directly from the liquid phase (such as liquid chromatography or capillary electrophoresis). It rarely disrupts the chemical nature of the analyst prior to mass analysis. Usually the sample is introduced into the ion source of the mass spectrometer through a capillary tube at atmospheric pressure. A high voltage is applied to the capillary, which creates an electric field in which the solvent rapidly evaporate and the charged droplets are divided into smaller droplets and eventually into individual charged molecules and enter the mass spectrometer. Positive and negative ESI ionization mode could be applied to analysis different type of lipid. Fatty acid can be analyzed using negative ionization mode, or using positive ionization mode after Li^+ ions are added. Glycerophosphatidylcholine could form cation molecular ions in mildly acidic solvents and could be analyzed using positive ionization mode, while glycerophosphatidylinositol is usually analyzed using negative ionization mode for the anion molecular ions of glycerophosphatidylinositol generate more product ions.[125] Glycerophosphatidylethanolamine can be ionized in positive mode to obtain the abundant ions corresponding to the neutral loss and in negative mode to obtain more information about the fatty acid species.[126] A rapid and highly sensitive shotgun lipidomic method has been developed to analyze lipids in biological samples based on the different ionization trend of different lipids in ESI. Lipids are divided into four species (anionic lipids, weak anionic lipids, neutral polar lipids, special lipids) based on their electrical propensities and analyzed using either negative or positive ionization mode.[4,99,127-130]

Lipid arrays have also been developed to identify and quantitate lipid species from crude lipid extract by ESI-MS flow injection analysis.[131,132] Besides lipid ESI-MS injection analysis, HPLC can also be connected with ESI-MS. In HPLC-ESI/MS, lipids are first separated using normal phase HPLC column or reverse phase HPLC column,[133-140] which can enhance the selectivity and sensitivity for lipid analysis. For phospholipids, a normal-phase column can separate most of the major classes of phospholipids, while reverse-phase column has even higher resolution, as reverse-phase column separates phospholipids based on their hydrophobicity, i.e., the length of their fatty acyl chains and the number of unsaturated bonds. Thus, normal-phase and reverse-phase could be combined in lipid analysis, in which normal-phase is used to collect fractions while reverse-phase is used to perform the accurate identification of molecular species.

Lipidomic Strategy in Cancer Biomarker Discovery

Even though we have discussed many cancer-related lipids above, however, most of those lipids were determined using the traditional NMR or biochemistry methods. Since there are many important lipid species, including sphingolipids and phospholipids, could not be analyzed by NMR and biochemistry methods due to their low sensitivity and other limitations, it is clear that with the newly developed lipidomic technology, many new and potential useful lipid biomarkers for cancer are waiting to be identified.

In cancer biomarker discovery, one of the most obvious applications of lipidomics is to compare the lipid profile under health and disease state, thus to identify those lipid species with altered expression. Unfortunately, to date, there are few reports about such systematic analysis for cancer lipid biomarker discovery, even though such effort is undertaken by several leading groups in lipidomic research. On the other hand, it is also acknowledged that a combination of the lipid biomarker with protein biomarker may provide a more accurate diagnosis. Gadomska et al has reported that the diagnosis of malignant ovarian could reach 97% using parameters including apolipoprotein and cholesterol.[16] Indeed, it is becoming increasingly evident that understanding the biology of lipids often requires the understanding of protein responses that are mediated by lipids. Since lipids could be analyzed by lipidomics, proteins are analyzed by proteomics and genes are analyzed by genomics, thus it is very important to combine lipidomics with proteomics and genomics for cancer biomarker study. The combined "omics" research forms the basis for systems biology, in which physiological events are studied at a system level. This type of systems biology research has already been applied in the study of a neural disease, neuronal ceroid lipofuscinoses (NCL).[141] NCL is a neural maladjustment disease which is caused by mutations of seven genes. Abnormal lipid metabolism has been reported in NCL as early as the 1960s[142] and 1980s,[143-145] with more changed lipid species were identified using lipidomics.[97,146,147] Similarly, combining lipidomics and proteomics, Fonteh et al also tried to identify disease markers in human cerebrospinal fluid from healthy and diseased people.[148] Such efforts will definitely enhance the existing knowledge of disease pathology and increase the likelihood of discovering specific markers and biochemical mechanisms of various diseases.

Conclusion

As a newly developed research area, lipidomics is still an open field inviting researchers who are interested in understanding the physiological events from a lipid angle. Lipid cancer biomarker discovery is of particular importance, as cancer remains one of the leading diseases for human death. However, so far such studies are rare and almost none useful lipid cancer biomarkers has been found. Therefore, more efforts and resources are needed for lipidomic studies.

Acknowledgements

This work is supported in part by grants from the National Natural Science Foundation of China (No. 30600220, 30771826 and 30872140), Science and Technology Department of Zhejiang Province, China (No. 2007F70031).

References

1. Wenk MR. The emerging field of lipidomics. Nat Rev Drug Discov 2005; 4:594-610.
2. Chisolm GM, Steinberg D. The oxidative modification hypothesis of atherogenesis: an overview. Free Radic Biol Med 2000; 28:1815-1826.
3. Watson AD. Thematic review series: systems biology approaches to metabolic and cardiovascular disorders. Lipidomics: a global approach to lipid analysis in biological systems. J Lipid Res 2006; 47:2101-2111.
4. Han X, Gross RW. Global analyses of cellular lipidomes directly from crude extracts of biological samples by ESI mass spectrometry: a bridge to lipidomics. J Lipid Res 2003; 44:1071-1079.
5. Lagarde M, Geloen A, Record M et al. Lipidomics is emerging. Biochim Biophys Acta 2003; 1634:61.
6. Ceschi M, Gutzwiller F, Moch H et al. Epidemiology and pathophysiology of obesity as cause of cancer. Swiss Med Wkly 2007; 137:50-56.
7. Calle EE, Kaaks R. Overweight, obesity and cancer: epidemiological evidence and proposed mechanisms. Nat Rev Cancer 2004; 4:579-591.
8. Fahy E, Subramaniam S, Brown HA et al. A comprehensive classification system for lipids. J Lipid Res 2005; 46:839-861.
9. Jiang J, Nilsson-Ehle P, Xu N. Influence of liver cancer on lipid and lipoprotein metabolism. Lipids Health Dis 2006; 5:4.
10. Riboni L, Ghidoni R, Sonnino S et al. Phospholipid content and composition of human meningiomas. Neurochem Pathol 1984; 2:171-188.
11. Szachowicz-Petelska B, Sulkowski S, Figaszewski ZA. Altered membrane free unsaturated fatty acid composition in human colorectal cancer tissue. Mol Cell Biochem 2007; 294:237-242.

12. Qadir MI, Malik SA. Plasma lipid profile in gynecologic cancers. Eur J Gynaecol Oncol 2008; 29:158-161.
13. Franky Dhaval S, Shilin Nandubhai S, Pankaj Manubhai S et al. Significance of alterations in plasma lipid profile levels in breast cancer. Integr Cancer Ther 2008; 7:33-41.
14. Patel PS, Shah MH, Jha FP et al. Alterations in plasma lipid profile patterns in head and neck cancer and oral precancerous conditions. Indian J Cancer 2004; 41:25-31.
15. Yamada K, Araki S, Tamura M et al. Relation of serum total cholesterol, serum triglycerides and fasting plasma glucose to colorectal carcinoma in situ. Int J Epidemiol 1998; 27:794-798.
16. Gadomska H, Grzechocinska B, Janecki J et al. Serum lipids concentration in women with benign and malignant ovarian tumours. Eur J Obstet Gynecol Reprod Biol 2005; 120:87-90.
17. Nandeesha H, Koner BC, Dorairajan LN. Altered insulin sensitivity, insulin secretion and lipid profile in nondiabetic prostate carcinoma. Acta Physiol Hung 2008; 95:97-105.
18. Meleh M, Pozlep B, Mlakar A et al. Determination of serum lysophosphatidic acid as a potential biomarker for ovarian cancer. J Chromatogr B Analyt Technol Biomed Life Sci 2007; 858:287-291.
19. Pozlep B, Meleh M, Kobal B et al. Use of lysophosphatidic acid in the management of benign and malignant ovarian tumors. Eur J Gynaecol Oncol 2007; 28:394-399.
20. Sutphen R, Xu Y, Wilbanks GD et al. Lysophospholipids are potential biomarkers of ovarian cancer. Cancer Epidemiol Biomarkers Prev 2004; 13:1185-1191.
21. Xiao Y, Chen Y, Kennedy AW et al. Evaluation of plasma lysophospholipids for diagnostic significance using electrospray ionization mass spectrometry (ESI-MS) analyses. Ann N Y Acad Sci 2000; 905:242-259.
22. Xu Y, Shen Z, Wiper DW et al. Lysophosphatidic acid as a potential biomarker for ovarian and other gynecologic cancers. JAMA 1998; 280:719-723.
23. Sullentrop F, Moka D, Neubauer S et al. 31P NMR spectroscopy of blood plasma: determination and quantification of phospholipid classes in patients with renal cell carcinoma. NMR Biomed 2002; 15:60-68.
24. Kuliszkiewicz-Janus M, Janus W, Baczynski S. Application of 31P NMR spectroscopy in clinical analysis of changes of serum phospholipids in leukemia, lymphoma and some other nonhaematological cancers. Anticancer Res 1996; 16:1587-1594.
25. Kuliszkiewicz-Janus M, Baczynski S. Chemotherapy-associated changes in 31P MRS spectra of sera from patients with multiple myeloma. NMR Biomed 1995; 8:127-132.
26. Zhao Z, Xiao Y, Elson P et al. Plasma lysophosphatidylcholine levels: potential biomarkers for colorectal cancer. J Clin Oncol 2007; 25:2696-2701.
27. Kuliszkiewicz-Janus M, Tuz MA, Baczynski S. Application of 31P MRS to the analysis of phospholipid changes in plasma of patients with acute leukemia. Biochim Biophys Acta 2005; 1737:11-15.
28. de Castro J, Rodriguez MC, Martinez-Zorzano VS et al. Erythrocyte and platelet phospholipid fatty acids as markers of advanced nonsmall cell lung cancer: comparison with serum levels of sialic acid, TPS and Cyfra 21-1. Cancer Invest 2008; 26:407-418.
29. Merchant TE, de Graaf PW, Minsky BD et al. Esophageal cancer phospholipid characterization by 31P NMR. NMR Biomed 1993; 6:187-193.
30. Merchant TE, Kasimos JN, Vroom T et al. Malignant breast tumor phospholipid profiles using (31)P magnetic resonance. Cancer Lett 2002; 176:159-167.
31. Eliyahu G, Kreizman T, Degani H. Phosphocholine as a biomarker of breast cancer: molecular and biochemical studies. Int J Cancer 2007; 120:1721-1730.
32. Nakagami K, Uchida T, Ohwada S et al. Increased choline kinase activity and elevated phosphocholine levels in human colon cancer. Jpn J Cancer Res 1999; 90:419-424.
33. Merchant TE, Kasimos JN, de Graaf PW et al. Phospholipid profiles of human colon cancer using 31P magnetic resonance spectroscopy. Int J Colorectal Dis 1991; 6:121-126.
34. Dobrzynska I, Szachowicz-Petelska B, Sulkowski S et al. Changes in electric charge and phospholipids composition in human colorectal cancer cells. Mol Cell Biochem 2005; 276:113-119.
35. Campanella R. Membrane lipids modifications in human gliomas of different degree of malignancy. J Neurosurg Sci 1992; 36:11-25.
36. Kal'nova N, Pal'mina NP. (Changes in plasma phospholipid composition and lipid antioxidant activity of breast tumor patients without and following radiation therapy). Vopr Med Khim 1981; 27:510-513.
37. Merchant TE, Diamantis PM, Lauwers G et al. Characterization of malignant colon tumors with 31P nuclear magnetic resonance phospholipid and phosphatic metabolite profiles. Cancer 1995; 76:1715-1723.
38. Khyshiktuev BS, Agapova Iu R, Zhilin IV et al (Phospholipid composition of various parts of the involved organ in lung cancer). Vopr Med Khim 1999; 45:350-354.
39. Dillehay DL, Webb SK, Schmelz EM et al. Dietary sphingomyelin inhibits 1,2-dimethylhydrazine-induced colon cancer in CF1 mice. J Nutr 1994; 124:615-620.

40. Schmelz EM, Dillehay DL, Webb SK et al. Sphingomyelin consumption suppresses aberrant colonic crypt foci and increases the proportion of adenomas versus adenocarcinomas in CF1 mice treated with 1,2-dimethylhydrazine: implications for dietary sphingolipids and colon carcinogenesis. Cancer Res 1996; 56:4936-4941.
41. Exon JH, South EH. Effects of sphingomyelin on aberrant colonic crypt foci development, colon crypt cell proliferation and immune function in an aging rat tumor model. Food Chem Toxicol 2003; 41:471-476.
42. Watanabe K, Matsubara T, Hakomori S. alpha-L-Fucopyranosylceramide, a novel glycolipid accumulated in some of the human colon tumors. J Biol Chem 1976; 251:2385-2387.
43. Ravindranath MH, Irie RF. Gangliosides as antigens of human melanoma. Cancer Treat Res 1988; 43:17-43.
44. Tsuchida T, Saxton RE, Morton DL et al. Gangliosides of human melanoma. J Natl Cancer Inst 1987; 78:45-54.
45. Ruf P, Jager M, Ellwart J et al. Two new trifunctional antibodies for the therapy of human malignant melanoma. Int J Cancer 2004; 108:725-732.
46. Tsuchida T, Saxton RE, Morton DL et al. Gangliosides of human melanoma. Cancer 1989; 63:1166-1174.
47. Saito S, Orikasa S, Ohyama C et al. Changes in glycolipids in human renal-cell carcinoma and their clinical significance. Int J Cancer 1991; 49:329-334.
48. Saito S, Orikasa S, Satoh M et al. Expression of globo-series gangliosides in human renal cell carcinoma. Jpn J Cancer Res 1997; 88:652-659.
49. Saito S, Nojiri H, Satoh M et al. Inverse relationship of expression between GM3 and globo-series ganglioside in human renal cell carcinoma. Tohoku J Exp Med 2000; 190:271-278.
50. Maruyama R, Saito S, Bilim V et al. High incidence of GalNAc disialosyl lactotetraosylceramide in metastatic renal cell carcinoma. Anticancer Res 2007; 27:4345-4350.
51. Arab S, Russel E, Chapman WB et al. Expression of the verotoxin receptor glycolipid, globotriaosylceramide, in ovarian hyperplasias. Oncol Res 1997; 9:553-563.
52. Andreasian GO, Malykh Ia N, Diatlovitskaia EV. (Ceramides and gangliosides in benign and malignant human ovarian tumors). Vopr Med Khim 1996; 42:248-253.
53. Satoh M, Fukushi Y, Kawamura J et al. Glycolipid expression in prostatic tissue and analysis of the antigen recognized by antiprostatic monoclonal antibody APG1. Urol Int 1992; 48:20-24.
54. Ohyama C, Orikasa S, Kawamura S et al (Immunohistochemical study of globotriaosyl ceramide (Gb3) in testicular tumors). Nippon Hinyokika Gakkai Zasshi 1993; 84:1308-1315.
55. Ohyama C, Orikasa S, Satoh M et al. Globotriaosyl ceramide glycolipid in seminoma: its clinicopathological importance in differentiation from testicular malignant lymphoma. J Urol 1992; 148:72-75.
56. Ohyama C, Orikasa S, Kawamura S et al. Galactosylgloboside expression in seminoma. Inverse correlation with metastatic potential. Cancer 1995; 76:1043-1050.
57. Sung CC, Pearl DK, Coons SW et al. Gangliosides as diagnostic markers of human astrocytomas and primitive neuroectodermal tumors. Cancer 1994; 74:3010-3022.
58. Pan XL, Izumi T, Yamada H et al. Ganglioside patterns in neuroepithelial tumors of childhood. Brain Dev 2000; 22:196-198.
59. Shinoura N, Dohi T, Kondo T et al. Ganglioside composition and its relation to clinical data in brain tumors. Neurosurgery 1992; 31:541-549.
60. Singh LP, Pearl DK, Franklin TK et al. Neutral glycolipid composition of primary human brain tumors. Mol Chem Neuropathol 1994; 21:241-257.
61. Sung CC, Pearl DK, Coons SW et al. Correlation of ganglioside patterns of primary brain tumors with survival. Cancer 1995; 75:851-859.
62. Comas TC, Tai T, Kimmel D et al. Immunohistochemical staining for ganglioside GD1b as a diagnostic and prognostic marker for primary human brain tumors. Neuro Oncol 1999; 1:261-267.
63. Popko B, Pearl DK, Walker DM et al. Molecular markers that identify human astrocytomas and oligodendrogliomas. J Neuropathol Exp Neurol 2002; 61:329-338.
64. Fredman P, von Holst H, Collins VP et al. Sialyllactotetraosylceramide, a ganglioside marker for human malignant gliomas. J Neurochem 1988; 50:912-919.
65. Jennemann R, Rodden A, Bauer BL et al. Glycosphingolipids of human gliomas. Cancer Res 1990; 50:7444-449.
66. Yates AJ, Franklin TK, McKinney P et al. Gangliosides and neutral glycolipids in ependymal, neuronal and primitive neuroectodermal tumors. J Mol Neurosci 1999; 12:111-121.
67. Yates AJ, Comas T, Scheithauer BW et al. Glycolipid markers of astrocytomas and oligodendrogliomas. J Neuropathol Exp Neurol 1999; 58:1250-1262.

68. Ono M, Handa K, Sonnino S et al. GM3 ganglioside inhibits CD9-facilitated haptotactic cell motility: coexpression of GM3 and CD9 is essential in the downregulation of tumor cell motility and malignancy. Biochemistry 2001; 40:6414-6421.
69. Yang J, Yu Y, Sun S et al. Ceramide and other sphingolipids in cellular responses. Cell Biochem Biophys 2004; 40:323-350.
70. Schenck M, Carpinteiro A, Grassme H et al. Ceramide: physiological and pathophysiological aspects. Arch Biochem Biophys 2007; 462:171-175.
71. Carpinteiro A, Dumitru C, Schenck M et al. Ceramide-induced cell death in malignant cells. Cancer Lett 2008; 264:1-210.
72. Simstein R, Burow M, Parker A et al. Apoptosis, chemoresistance and breast cancer: insights from the MCF-7 cell model system. Exp Biol Med (Maywood) 2003; 228:995-1003.
73. Tan X, Zhang Y, Jiang B et al. Changes in ceramide levels upon catechins-induced apoptosis in LoVo cells. Life Sci 2002; 70:2023-2029.
74. Zhang XF, Li BX, Dong CY et al. Apoptosis of human colon carcinoma HT-29 cells induced by ceramide. World J Gastroenterol 2006; 12:3581-3584.
75. Ahn EH, Schroeder JJ. Sphingoid bases and ceramide induce apoptosis in HT-29 and HCT-116 human colon cancer cells. Exp Biol Med (Maywood) 2002; 227:345-353.
76. Sumitomo M, Ohba M, Asakuma J et al. Protein kinase Cdelta amplifies ceramide formation via mitochondrial signaling in prostate cancer cells. J Clin Invest 2002; 109:827-836.
77. Van Brocklyn JR. Sphingolipid signaling pathways as potential therapeutic targets in gliomas. Mini Rev Med Chem 2007; 7:984-990.
78. Riboni L, Campanella R, Bassi R et al. Ceramide levels are inversely associated with malignant progression of human glial tumors. Glia 2002; 39:105-113.
79. Reynolds CP, Maurer BJ, Kolesnick RN. Ceramide synthesis and metabolism as a target for cancer therapy. Cancer Lett 2004; 206:169-180.
80. Lin CF, Chen CL, Lin YS. Ceramide in apoptotic signaling and anticancer therapy. Curr Med Chem 2006; 13:1609-1616.
81. Savtchouk IA, Mattie FJ, Ollis AA. Ceramide: from embryos to tumors. Sci STKE 2007; 2007:jc1.
82. Huwiler A, Pfeilschifter J. Altering the sphingosine-1-phosphate/ceramide balance: a promising approach for tumor therapy. Curr Pharm Des 2006; 12:4625-4635.
83. Huwiler A, Zangemeister-Wittke U. Targeting the conversion of ceramide to sphingosine 1-phosphate as a novel strategy for cancer therapy. Crit Rev Oncol Hematol 2007; 63:150-159.
84. Sanchez T, Hla T. Structural and functional characteristics of S1P receptors. J Cell Biochem 2004; 92:913-922.
85. Spiegel S, Cuvillier O, Edsall LC et al. Sphingosine-1-phosphate in cell growth and cell death. Ann N Y Acad Sci 1998; 845:11-18.
86. Spiegel S, Cuvillier O, Edsall L et al. Roles of sphingosine-1-phosphate in cell growth, differentiation and death. Biochemistry (Mosc) 1998; 63:69-73.
87. Bassi R, Anelli V, Giussani P et al. Sphingosine-1-phosphate is released by cerebellar astrocytes in response to bFGF and induces astrocyte proliferation through Gi-protein-coupled receptors. Glia 2006; 53:621-630.
88. Cuvillier O, Rosenthal DS, Smulson ME et al. Sphingosine 1-phosphate inhibits activation of caspases that cleave poly(ADP-ribose) polymerase and lamins during Fas- and ceramide-mediated apoptosis in Jurkat T-lymphocytes. J Biol Chem 1998; 273:2910-2916.
89. Cuvillier O, Pirianov G, Kleuser B et al. Suppression of ceramide-mediated programmed cell death by sphingosine-1-phosphate. Nature 1996; 381:800-803.
90. Castillo SS, Teegarden D. Sphingosine-1-phosphate inhibition of apoptosis requires mitogen-activated protein kinase phosphatase-1 in mouse fibroblast C3H10T 1/2 cells. J Nutr 2003; 133:3343-3349.
91. Spiegel S, Merrill AH Jr. Sphingolipid metabolism and cell growth regulation. FASEB J 1996; 10:1388-1397.
92. Spiegel S, Olivera A, Zhang H et al. Sphingosine-1-phosphate, a novel second messenger involved in cell growth regulation and signal transduction, affects growth and invasiveness of human breast cancer cells. Breast Cancer Res Treat 1994; 31:337-348.
93. Yamamura S, Sadahira Y, Ruan F et al. Sphingosine-1-phosphate inhibits actin nucleation and pseudopodium formation to control cell motility of mouse melanoma cells. FEBS Lett 1996; 382:193-197.
94. Liu X, Elojeimy S, Turner LS et al. Acid ceramidase inhibition: a novel target for cancer therapy. Front Biosci 2008; 13:2293-2298.
95. Folch J, Lees M, Sloane Stanley GH. A simple method for the isolation and purification of total lipides from animal tissues. J Biol Chem 1957; 226:497-509.
96. Bligh EG, Dyer WJ. A rapid method of total lipid extraction and purification. Can J Biochem Physiol 1959; 37:911-917.

97. Hermansson M, Kakela R, Berghall M et al. Mass spectrometric analysis reveals changes in phospholipid, neutral sphingolipid and sulfatide molecular species in progressive epilepsy with mental retardation, EPMR, brain: a case study. J Neurochem 2005; 95:609-617.

98. Saunders RD, Horrocks LA. Simultaneous extraction and preparation for high-performance liquid chromatography of prostaglandins and phospholipids. Anal Biochem 1984; 143:71-75.

99. Han X, Yang J, Cheng H et al. Toward fingerprinting cellular lipidomes directly from biological samples by two-dimensional electrospray ionization mass spectrometry. Anal Biochem 2004; 330:317-331.

100. Fujiwaki T, Tasaka M, Takahashi N et al. Quantitative evaluation of sphingolipids using delayed extraction matrix-assisted laser desorption ionization time-of-flight mass spectrometry with sphingosylphosphorylcholine as an internal standard. Practical application to cardiac valves from a patient with Fabry disease. J Chromatogr B Analyt Technol Biomed Life Sci 2006; 832:97-102.

101. Avigan J, Goodman DS, Steinberg D. Thin-Layer Chromatography of Sterols and Steroids. J Lipid Res 1963; 4:100-101.

102. Powell WS. Rapid extraction of oxygenated metabolites of arachidonic acid from biological samples using octadecylsilyl silica. Prostaglandins 1980; 20:947-957.

103. Powell WS. Rapid extraction of arachidonic acid metabolites from biological samples using octadecylsilyl silica. Methods Enzymol 1982; 86:467-477.

104. Kaluzny MA, Duncan LA, Merritt MV et al. Rapid separation of lipid classes in high yield and purity using bonded phase columns. J Lipid Res 1985; 26:135-140.

105. Hawkins DJ, Kuhn H, Petty EH et al. Resolution of enantiomers of hydroxyeicosatetraenoate derivatives by chiral phase high-pressure liquid chromatography. Anal Biochem 1988; 173:456-462.

106. Malavolta M, Bocci F, Boselli E et al. Normal phase liquid chromatography-electrospray ionization tandem mass spectrometry analysis of phospholipid molecular species in blood mononuclear cells: application to cystic fibrosis. J Chromatogr B Analyt Technol Biomed Life Sci 2004; 810:173-186.

107. Lesnefsky EJ, Stoll MS, Minkler PE et al. Separation and quantitation of phospholipids and lysophospholipids by high-performance liquid chromatography. Anal Biochem 2000; 285:246-254.

108. McHowat J, Jones JH, Creer MH. Gradient elution reversed-phase chromatographic isolation of individual glycerophospholipid molecular species. J Chromatogr B Biomed Sci Appl 1997; 702:21-32.

109. Nakamura T, Bratton DL, Murphy RC. Analysis of epoxyeicosatrienoic and monohydroxyeicosatetraenoic acids esterified to phospholipids in human red blood cells by electrospray tandem mass spectrometry. J Mass Spectrom 1997; 32:888-896.

110. Weintraub ST, Lear CS, Pinckard RN. Analysis of platelet-activating factor by GC-MS after direct derivatization with pentafluorobenzoyl chloride and heptafluorobutyric anhydride. J Lipid Res 1990; 31:719-725.

111. Balazy M, Braquet P, Bazan NG. Determination of platelet-activating factor and alkyl-ether phospholipids by gas chromatography-mass spectrometry via direct derivatization. Anal Biochem 1991; 196:1-10.

112. Wang Y, Krull IS, Liu C et al. Derivatization of phospholipids. J Chromatogr B Analyt Technol Biomed Life Sci 2003; 793:3-14.

113. Schiller J, Suss R, Arnhold J et al. Matrix-assisted laser desorption and ionization time-of-flight (MALDI-TOF) mass spectrometry in lipid and phospholipid research. Prog Lipid Res 2004; 43:449-488.

114. Fujiwaki T, Yamaguchi S, Tasaka M et al. Evaluation of sphingolipids in vitreous bodies from a patient with Gaucher disease, using delayed extraction matrix-assisted laser desorption ionization time-of-flight mass spectrometry. J Chromatogr B Analyt Technol Biomed Life Sci 2004; 806:47-51.

115. Fujiwaki T, Yamaguchi S, Tasaka M et al. Application of delayed extraction-matrix-assisted laser desorption ionization time-of-flight mass spectrometry for analysis of sphingolipids in pericardial fluid, peritoneal fluid and serum from Gaucher disease patients. J Chromatogr B Analyt Technol Biomed Life Sci 2002; 776:115-123.

116. Fujiwaki T, Yamaguchi S, Sukegawa K et al. Application of delayed extraction matrix-assisted laser desorption ionization time-of-flight mass spectrometry for analysis of sphingolipids in cultured skin fibroblasts from sphingolipidosis patients. Brain Dev 2002; 24:170-173.

117. Fujiwaki T, Tasaka M, Yamaguchi S. Quantitative evaluation of sphingomyelin and glucosylceramide using matrix-assisted laser desorption ionization time-of-flight mass spectrometry with sphingosylphosphorylcholine as an internal standard Practical application to tissues from patients with Niemann-Pick disease types A and C and Gaucher disease. J Chromatogr B Analyt Technol Biomed Life Sci 2008; 870:170-176.

118. Huang Y, Shen J, Wang T et al. A lipidomic study of the effects of N-methyl-N'-nitro-N-nitrosoguanidine on sphingomyelin metabolism. Acta Biochim Biophys Sin (Shanghai) 2005; 37:515-524.

119. Huang Y, Yang J, Shen J et al. Sphingolipids are involved in N-methyl-N'-nitro-N-nitrosoguanidine-induced epidermal growth factor receptor clustering. Biochem Biophys Res Commun 2005; 330:430-438.

120. Ma X, Liu G, Wang S et al. Evaluation of sphingolipids changes in brain tissues of rats with pentylenetetrazol-induced kindled seizures using MALDI-TOF-MS. J Chromatogr B Analyt Technol Biomed Life Sci 2007; 859:170-177.

121. Jackson SN, Wang HY, Woods AS. Direct profiling of lipid distribution in brain tissue using MALDI-TOFMS. Anal Chem 2005; 77:4523-4527.

122. Jackson SN, Wang HY, Woods AS. In situ structural characterization of phosphatidylcholines in brain tissue using MALDI-MS/MS. J Am Soc Mass Spectrom 2005; 16:2052-2056.

123. Jackson SN, Wang HY, Woods AS et al. Direct tissue analysis of phospholipids in rat brain using MALDI-TOFMS and MALDI-ion mobility-TOFMS. J Am Soc Mass Spectrom 2005; 16:133-138.

124. Jackson SN, Wang HY, Woods AS. In situ structural characterization of glycerophospholipids and sulfatides in brain tissue using MALDI-MS/MS. J Am Soc Mass Spectrom 2007; 18:17-26.

125. Pulfer M, Murphy RC. Electrospray mass spectrometry of phospholipids. Mass Spectrom Rev 2003; 22:332-364.

126. Larsen A, Uran S, Jacobsen PB et al. Collision-induced dissociation of glycero phospholipids using electrospray ion-trap mass spectrometry. Rapid Commun Mass Spectrom 2001; 15:2393-2398.

127. Han X, Gross RW. Electrospray ionization mass spectroscopic analysis of human erythrocyte plasma membrane phospholipids. Proc Natl Acad Sci USA 1994; 91:10635-10639.

128. Han X, Gross RW. Shotgun lipidomics: multidimensional MS analysis of cellular lipidomes. Expert Rev Proteomics 2005; 2:253-264.

129. Han X, Gross RW. Shotgun lipidomics: electrospray ionization mass spectrometric analysis and quantitation of cellular lipidomes directly from crude extracts of biological samples. Mass Spectrom Rev 2005; 24:367-412.

130. Han X, Yang J, Cheng H et al. Shotgun lipidomics identifies cardiolipin depletion in diabetic myocardium linking altered substrate utilization with mitochondrial dysfunction. Biochemistry 2005; 44:16684-16694.

131. Ivanova PT, Milne SB, Forrester JS et al. LIPID arrays: new tools in the understanding of membrane dynamics and lipid signaling. Mol Interv 2004; 4:86-96.

132. Milne S, Ivanova P, Forrester J et al. Lipidomics: an analysis of cellular lipids by ESI-MS. Methods 2006; 39:92-103.

133. Uran S, Larsen A, Jacobsen PB et al. Analysis of phospholipid species in human blood using normal-phase liquid chromatography coupled with electrospray ionization ion-trap tandem mass spectrometry. J Chromatogr B Biomed Sci Appl 2001; 758:265-275.

134. Houjou T, Yamatani K, Imagawa M et al. A shotgun tandem mass spectrometric analysis of phospholipids with normal-phase and/or reverse-phase liquid chromatography/electrospray ionization mass spectrometry. Rapid Commun Mass Spectrom 2005; 19:654-666.

135. Rainville PD, Stumpf CL, Shockcor JP et al. Novel application of reversed-phase UPLC-oaTOF-MS for lipid analysis in complex biological mixtures: a new tool for lipidomics. J Proteome Res 2007; 6:552-558.

136. Yoo HH, Son J, Kim DH. Liquid chromatography-tandem mass spectrometric determination of ceramides and related lipid species in cellular extracts. J Chromatogr B Analyt Technol Biomed Life Sci 2006; 843:327-333.

137. Bielawski J, Szulc ZM, Hannun YA et al. Simultaneous quantitative analysis of bioactive sphingolipids by high-performance liquid chromatography-tandem mass spectrometry. Methods 2006; 39:82-91.

138. Guan XL, He X, Ong WY et al. Non-targeted profiling of lipids during kainate-induced neuronal injury. FASEB J 2006; 20:1152-1161.

139. Ogiso H, Suzuki T, Taguchi R. Development of a reverse-phase liquid chromatography electrospray ionization mass spectrometry method for lipidomics, improving detection of phosphatidic acid and phosphatidylserine. Anal Biochem 2008; 375:124-131.

140. Retra K, Bleijerveld OB, van Gestel RA et al. A simple and universal method for the separation and identification of phospholipid molecular species. Rapid Commun Mass Spectrom 2008; 22:1853-1862.

141. Jalanko A, Tyynela J, Peltonen L. From genes to systems: new global strategies for the characterization of NCL biology. Biochim Biophys Acta 2006; 1762:934-944.

142. Hagberg B, Sourander P, Svennerholm L. Late infantile progressive encephalopathy with disturbed poly-unsaturated fat metabolism. Acta Paediatr Scand 1968; 57:495-499.

143. Svennerholm L, Fredman P, Jungbjer B et al. Large alterations in ganglioside and neutral glycosphingolipid patterns in brains from cases with infantile neuronal ceroid lipofuscinosis/polyunsaturated fatty acid lipidosis. J Neurochem 1987; 49:1772-1783.

144. Palmer DN, Husbands DR, Jolly RD. Phospholipid fatty acids in brains of normal sheep and sheep with ceroid-lipofuscinosis. Biochim Biophys Acta 1985; 834:159-163.

145. Wolfe LS, Gauthier S, Haltia M et al. Dolichol and dolichyl phosphate in the neuronal ceroid-lipofuscinoses and other diseases. Am J Med Genet Suppl 1988; 5:233-242.

146. Kakela R, Somerharju P, Tyynela J. Analysis of phospholipid molecular species in brains from patients with infantile and juvenile neuronal-ceroid lipofuscinosis using liquid chromatography-electrospray ionization mass spectrometry. J Neurochem 2003; 84:1051-1065.

147. Granier LA, Langley K, Leray C et al. Phospholipid composition in late infantile neuronal ceroid lipofuscinosis. Eur J Clin Invest 2000; 30:1011-1017.

148. Fonteh AN, Harrington RJ, Huhmer AF et al. Identification of disease markers in human cerebrospinal fluid using lipidomic and proteomic methods. Dis Markers 2006; 22:39-64.

CHAPTER 10

Bioinformatics in Cancer Biomarker Discovery

Yee Leng Yap* and Xuewu Zhang

Abstract

Many bioinformatics methods analyzing data generated from "omics" platforms have been developed for two main objectives: to identify reliable cancer biomarkers for early cancer diagnosis and to discover molecular targets for designed therapeutic intervention. A review of the bioinformatics literature identifying cancer biomarkers from various platforms is presented in this chapter. The existing problems, solutions and future prospects have been briefly discussed. With the rapid advancement of bioinformatics techniques and robust "omics" platforms, the biomarker-based applications in clinical oncology will, in the foreseeable future, gradually enter into the mainstream clinics as an important addition to current clinical strategies.

Introduction to Bioinformatics and Cancer Biomarker Discovery

The primary goal of bioinformatics in cancer biomarker discovery is to increase our understanding of cancer development processes. What sets it apart from other approaches is its focus on developing and applying computationally intensive techniques (data mining, machine learning algorithms involving applied mathematics, informatics, statistics, computer science, artificial intelligence, chemistry and biochemistry) to achieve this goal.[1-3] Cancer is generally a genetic disease.[4] More than 11 million people are diagnosed with cancer annually. It is estimated that there will be 16 million new cases every year by 2020. From a total of 58 million deaths worldwide in 2005, cancer fatality accounts for 7.6 million (~13%). To make the matter worse, global deaths resulted from cancers are projected to continue rising.[5] Some of the best available strategies to fight cancer include primary chemoprevention, early diagnosis and anticancer therapy. Because most of the mechanisms for cancer initiation and progression remain illusive, no robust strategy is clinically effective for primary chemoprevention and removal of cancer cells specifically. Although many drugs are designed against cancers, the death rates for the most prevailing cancers are not reduced. Thus, much attention is paid to cancer biomarker discovery for early cancer diagnosis.[6]

To date, effective cancer screening is available for a few types of cancer, including: (1) Pap smear for cervical cancer detection,[7] (2) mammography for breast cancer detection,[8] (3) fecal occult blood testing (FOBT) for colorectal cancer detection,[9] (4) pepsinogen I (PGI) and gastrin-17 (G17) test for gastric cancer detection,[10] (5) prostate specific antigen (PSA) for prostate cancer detection.[11] In addition, some other FDA-approved[12] cancer biomarkers, such as carcinoembryonic antigen (CEA), cancer antigen 125 (CA125), cancer antigen 153 (CA153), alpha-fetoprotein (α-FP) are also used for cancer diagnostics. Screening and diagnostic tests are typically evaluated by their sensitivity and specificity.[13] Sensitivity is the fraction of disease cases that are correctly identified as diseased; specificity is the fraction of nondisease cases that are correctly identified as

*Corresponding Author: Yee Leng Yap—Davos Life Science Pte. Ltd., 11 Biopolis Way, The Helios #07-03, 138667 Singapore. Email: daniel.yap@davoslife.com

Omics Technologies in Cancer Biomarker Discovery, edited by Xuewu Zhang.
©2011 Landes Bioscience.

nondiseased. However, the accuracy levels of the above screening methods are far from ideal. The sensitivity and specificity of Pap smear is 73% and 63%[14] respectively. Although the sensitivity of mammography is approximately 90%, only 15-34% of the positive cases from mammography are actual malignant tissue.[14] FOBT has a high specificity of 96-98%, but its sensitivity is ~40%.[15] The false negatives delay the diagnosis and treatment of cancer. At present, majority of cancer patients are diagnosed as having late stage cancer. Such as, 72% of lung cancer patients, 57% of colorectal cancer patients and 34% of breast cancer patients in the United States are diagnosed at late stage. If these cancers are diagnosed at early stage, the 5-year survival rate exceeds 85%.[16] For example, in ovarian cancer, most women with this cancer have advanced disease at diagnosis, the 5-year survival rate for ovarian cancer patients at the advanced stage is 35%, meanwhile the 5-year survival rate for ovarian cancer patients at the Stage I is exceed 90%. Similarly, when breast cancer is detected at the advanced stage with metastasis from the original organ, the survival rate of the patients drops below 23%. This is in contrast to the case when the survival rate of the patient increases to 97% when cancer is discovered at early stage. Another example is that the survival rate of prostate cancer patient improves from 34% when the cancer is detected at the advanced stage, to >98% at the early stage.[5] Thus, it is very important to develop highly sensitive and specific biomarkers to diagnose cancer at early stages, which directly affect mortality. In this chapter, we summarized the bioinformatics advances relating to the discovery of cancer biomarkers using data generated from "omics" platforms. The potential problems, solutions and prospects are also discussed.

Cancer Biomarker Discovery Using Various 'Omics' Platforms

Current bioinformatics efforts are focusing on biomarker discovery. The validated biomarkers can either be used for the earlier diagnosis of cancer, or become molecular target for designed therapeutic intervention. The large scale datasets being analyzed are usually generated from various "omics" technologies, such as genomics,[17] transcriptomics,[18] proteomics[19] and metabonomics,[20] which are four functional parts of organisms (DNA, RNA, protein, metabolites). The general strategy for discovering potential cancer biomarkers is to identify the differentially expressed genes, transcripts, proteins or metabolites through comparing molecular profiles between benign tissue and cancer tissue. The comparison usually involves statistical hypothesis testing followed by some independent cross validation strategies to indicate significance of the results.[21] Some potential cancer biomarkers identified by bioinformatics techniques are summarized in Table 1.

Briefly, genomics covers the effort to decipher the complete DNA sequence of organisms and fine-scale single-gene genetic mapping. Usually, attacks by mutagens such as ionizing radiations, DNA base analogs, intercalating agents, alkylating agents result in DNA mutations that potentially cause cancer formation[22] and affect therapeutic intervention. For example, somatic mutations of the epidermal growth factor receptor gene (EGFR) were found to be responsible for the overall response of EGFR kinase inhibitor therapy (Gefitinib). In one report where EGFR was sequenced in nonsmall cell lung cancer and matched normal tissue, treatment with the Gefitinib causes tumor regression in some patients with specific mutation in EGFR.[23] Another example will be the inherited defects in DNA repair genes, such as BRCA1, BRCA2 as found to be strongly implicated in breast cancer incidences.[24] Separately, DNA methylations catalyzed by the enzyme DNA methyltransferase have a major impact on gene activity and cancer development. Progress in past years suggests that hypermethylation might be caused by an aberrant reactivation of protein complexes, which normally mediate the silencing of genes by DNA methylation and chromatin remodeling during cell differentiation.[25] In some special case, DNA hypomethylation has also caused genomic instability and increase cancer risk. An example, a small, but significantly higher amount of global hypomethylation in the leucocytes of patients with cancer compared with those with noncancerous diseases.[17] To discover these lethal mutations, new sequencing methods (SNP array,[26] In vitro clonal amplification,[27] SAGE[28]) may have advantages in terms of efficiency or accuracy.[29] The resulted large amount of genotyping data will be subjected to bioinformatics genome-wide association study to identify genetic associations with disease traits[4,26] (Table 1).

Transcriptomics employ an array of molecular probes complementary to specific target mRNA sequences, which is fixed onto a solid phase. The probes may be synthesized directly onto the solid phase by a photolithographic technique or printed PCR-products/oligonucleotides. RNA is extracted from tissues of interest, labeled with a detectable dye and hybridized to the microarray chip. Confocal laser scanning is used to render the images and the relative fluorescence intensity of each gene-specific probe represents the level of expression of the particular gene. Microarray technology allows a genome-wide screening of gene expression due to the use of high density of probes.[30] Since its introduction, microarray technology has been used to identify biomarkers associated with diseases and the patterns of gene expression could be used to classify types of tumors and predict the outcome. Numerous reports have demonstrated, with various bioinformatics models, the potential power of expression profiling for molecular diagnosis of human cancers (Table 1).

Proteomics refers to the large-scale study of large proteins, particularly their structures and functions. Among "omics" platforms, the fast, sensitive and robust mass spectrometry (MS) are emerging as a key technology for biomarkers discovery and clinical cancer diagnosis.[31] The previous protein markers discovery techniques such as RIA and ELISA were limited by their requirement of prior knowledge of the molecule. Two-dimensional polyacrylamide gel electrophoresis (2D-Gel or PAGE) provides an alternative approach for proteomics analysis, but 2D-Gel has inferior mass resolution and sensitivity, undetectable for the mass range <20 kD. MS instruments with two different ionization techniques are mostly used for high resolution proteomics experiments. One is electrospray ionization (ESI) MS. The second is laser desorption and ionization MS, including matrix-assisted laser desorption and ionization (MALDI) with a time-of-flight MS and surface-enhanced laser desorption and ionization (SELDI) with a time-of-flight MS.[32] The detector plate records the intensity of the signal at a given m/z value and a spectrum is generated.[33] The currently available MS datasets are presented in two formats: (1) a panel of differentially displayed peaks from normal and cancer. Although individual peak is not identified, this will not limit its utility for medical diagnostics since cancer diagnostic is a problem of prediction rather than of aetiology,[34] (2) the discriminating peaks are at least further fragmented by the use of tandem mass spectrometry (MS/MS). However, the important bioinformatics challenges for proteomic studies lie in peptide/protein identification using variety of search algorithms and protein databases. Some successful bioinformatics strategies used to discover potential biomarkers are tabulated in Table 1.

Metabonomics represents large-scale study into the end stage of all molecular events. The number of metabolites is around 5,000, compared with 1,000,000 estimated proteins and over 30,000 human genes.[35] In addition, it is now clear that investigating RNA/protein changes associated with carcinogenesis does not necessarily lead to functional understanding. Rather, changes in the metabonome, as the downstream product of gene expression and protein translation, are expected to be amplified and easily detectable.[36] In order to obtain a systematic elucidation of cancer cell dysfunction, the metabolic consequences of gene products must be investigated. Generally, there are two names "metabolomics" and "metabonomics". Metabolomics is the measurement of all metabolites concentrations in cells and tissues and metabonomics is the quantitative measurement of metabolic responses of multicellular systems to pathophysiological stimuli or genetic modification.[37] Several methods can generate metabolic signatures of biomaterials, including nuclear magnetic resonance (NMR), gas chromatography-MS (GC-MS), liquid chromatography-MS (LC-MS), capillary electrophoresis-MS (CE-MS). New advancements in CE-MS and GC-MS enable quantitative analysis of a relatively large number of metabolites.[38,39] In Table 1, we highlighted several bioinformatics studies reporting how metabonomic data can be analyzed for identification of cancer biomarkers.

Bioinformatics Solutions and Challenges in Cancer Biomarker Discovery

All "omics" technologies rely on analytical chemistry and result in complex multivariate datasets. In order to extract disease information that is of diagnostic or prognostic value, a variety of bioinformatics tools are required for interpreting the changes in these "omics" datasets.

Table 1. Examples of bioinformatics applications in cancer biomarker discovery

No.	Platform	Findings	Ref
1	Genomics	Prostate cancer: Rapid progression of prostate cancer in men with a BRCA2 mutation	40
2		Breast cancer: BRCA1, BRCA2 are strongly implicated in breast cancer incidences	41
3		Colon cancer: TP53 mutation is useful for the prediction of disease-free survival in adjuvant-treated Stage III colon cancers	42
4		Lung cancer: Somatic EGFR mutations are responsible for the overall response of EGFR kinase inhibitor therapy	43
5		Pancreatic cancer: Palladin mutation causes familial pancreatic cancer	44
6		Ovarian cancer: BRCA1/BRCA2 mutation prevalence and clinical characteristics of ovarian cancer cases	45
7		Gastric cancer: PCR-SSCP-DNA sequencing method in detecting PTEN gene mutation in gastric cancer	46
8		Bladder cancer: DNA hypomethylation has caused genomic instability and increase cancer risk	17
9	Transcriptomics	Prostate cancer: Integrative microarray analysis of pathways dysregulated in metastatic prostate cancer	47
10		Breast cancer: Biomarker signatures identified for predicting recurrence	48, 49
11		Colon cancer: EGFR, c-MET, beta-catenin and p53 expression as prognostic indicators in colon cancer	50
12		Lung cancer: Gene expression ratio is an accurate technique for distinguishing between MPM and lung cancer	51
13		Pancreatic cancer: Meta-analysis revealed that the pancreatic cancer gene expression shared a significant number of dysregulated genes associated with the cell adhesion-mediated drug resistance pathway	52
14		Ovarian cancer: Microarray identifies Prostasin, a potential serum marker for ovarian cancer	53
15		Gastric cancer: Gene signatures can be used to diagnose cancer and its histological subtypes	54
16	Proteomics	Prostate cancer: Identification of precursor forms of free prostate-specific antigen in serum of prostate cancer patients	55
17		Breast cancer: MS protein profiling identifies ubiquitin and ferritin light chain as prognostic biomarkers in breast cancers	56
18		Colon cancer: MS revealed abnormal distribution of phospholipids in colon cancer liver metastasis	57
19		Lung cancer: An algorithm based on m/z features was developed based on outcomes after EGFR inhibitors therapy	58

continued on next page

Table 1. Continued

No.	Platform	Findings	Ref
20		Pancreatic cancer: Ensemble classifiers always outperform single decision tree classifier in classifying unknown sample	59
21		Ovarian cancer: Bayesian neural network approaches to ovarian cancer identification	60
22		Gastric cancer: A peptide fragment (fibrinopeptide A) was found to be highly elevated in cancer sera. The mean logarithmic concentrations of serum fibrinopeptide A levels were significantly higher in cancer patients and high-risk individuals.	61
23	Metabonomics	Prostate cancer: Cyclooxygenase/arachidonic acid metabolites were discovered in invasive human prostate cancer cells	62
24		Breast cancer: Dietary daidzein had a slight but significant stimulatory effect on MCF-7 tumor growth in mice	63
25		Lung cancer: Drug pharmacokinetics in patients with advanced nonsmall-cell lung cancer in a Phase II trial with gemcitabine administered as a fixed dose-rate infusion was investigated	64
26		Ovarian cancer: A protein (m/z approximately 17,400) with higher peak intensities in cancer patients than in benign conditions was identified as eosinophil-derived neurotoxin	65

In bioinformatics research, many discriminatory patterns can be generated and used for cancer diagnosis. For example, the gene patterns generated by RNA expression levels of thousands of genes from a cancer sample can be used to predict a patient's prognosis or response to therapy; the protein/peptide patterns generated by thousands of peaks from MS of serum can be used for diagnosis of persons with and without a cancer. A number of approaches were employed to discover these discriminatory patterns. For genomics, Text/Databases mining methods,[66] Sequence alignment strategies,[67] Frequency of occurrences,[68] Spectral clustering,[69] Information theory (entropy),[70] Hidden Markov model,[71] Bayesian model[72] and Correlation study.[73] For transcriptomics, diverse approaches are used to conduct diagnostic and prognostic prediction for cancer patients based on gene expression profiles: Genetic algorithm[74] and K-nearest neighbor method,[75] Fisher's linear discriminant analysis,[76] Recursive portioning method,[77] Two-way clustering,[78] Principle component analysis,[79] Partial least squares,[80] Pairwise gene expression ratio,[81] Independent component analysis,[82] Gene ontology method.[83] Different algorithms for gene expression analysis of cancer patients have been compared.[84] For proteomics, Genetic algorithms,[85] Decision-tree analysis,[86] Artificial neural network,[87] Linear or Quadratic discriminant analysis,[88] Support vector machines,[89] Self-organizing maps.[90] Evaluation of various methods is available in a methodological review.[91] For metabonomics, PCA method was used for diagnosis of liver cancer.[92] Theoretically no one particular classification method should be better than others for "omics" data analysis. The exploratory use of different algorithms and classifiers should be encouraged.[93]

Regardless of the approach used, the overfitting of data may be a problem. Overfitting refers to the situation where the number of parameters of the model is large compared to the size of training samples. This leads to the consequence that the model fits the training data arbitrarily, but performs poorly for independent datasets. Frequently cross-validation strategy is used to avoid overfitting. The cross-validation procedure can produce a nearly unbiased estimate of the true error rate of the classification procedure, but it must be emphasized that the cross-validation procedure should be properly performed in the entire model-building process.[94] Moreover, it is essential that additional validation step be performed on independent datasets, especially for class prediction studies with small sample sizes. There are at least two reasons for this: (1) a large variance for small sample sizes exists, although cross-validated error estimates are nearly unbiased;[95] (2) the training dataset used to build the predictor may not accurately represent the larger population.[94] Unfortunately, only ~10% of reports about microarray research conducted an independent validation.[93] The remaining reports either used a small validation set or a nonindependent validation set.[49]

Another major concern is that different bioinformatics analyses generate different predictive patterns. For example, using the same colon cancer microarray dataset, different data analysis methods produced different predictive biomarker lists and signature patterns.[77,78,82] Similarly in proteomics, different bioinformatics tools identified different discriminatory peaks.[86,96] To explain these discrepancies, Ein-Dor et al showed that the main problem was the small number of samples which hinder the identification of true gene signatures.[97,98] Indeed, if the dataset does not represent the underlying probability distribution of the population of interest, then even the most sophisticated sampling techniques will end up with an extremely biased result.[99] Xu et al[100] applied the top-scoring pair classifier (TSP) to three different prostate cancer datasets and identified a pair of robust biomarkers. Their results demonstrated that different datasets generate distinct TSPs when the sample size is small. However, as the sample size exceeded 135, the TSPs stabilize at their gene pair. Some other methods for determining optimum sample sizes in microarray studies have been reported.[101,102] Recently, Ein-Dor et al[97] introduced a new mathematical method, probably approximately correct (PAC) sorting, for calculating the number of arrays that are needed to achieve any desired level of reproducibility. The results demonstrated that thousands of samples are needed to generate a robust gene list for predicting outcome in cancer. Other estimated a less pessimistic finding of <20 arrays.[103] Unfortunately, no study is performed to estimate the efficient sample sizes for other "omics" research.

Integration of datasets obtained from different laboratories is a common strategy to increase sample size. For example, in proteomics, the Plasma Proteome Project has collected MS/MS datasets from 35 laboratories.[104] In order to maximize thorough datamining, utilization and warehousing of microarray data, a web-based cancer microarray platform, ONCOMINE, was developed.[105] It contains >100 gene expression datasets comprising nearly 48 million gene expression measurements. Other microarray databases include ArrayExpress,[106] SMD,[107] GEO.[108] Based on this, common transcriptional profile that is universally activated in most cancer types, a cancer meta-signature, was characterized.[109] This is not surprising as all cancer types share the common features of unregulated cell proliferation. More importantly, universal overexpression suggests that these genes may serve as attractive therapeutic targets with broad clinical application. For example, TOP2A, which is present in 10 types of cancer have been targeted therapeutically with some degree of success. Furthermore, the validated meta-signature genes coupled with KEGG pathway database[111] query enable discovery of important metabolic pathways underlying mechanisms of carcinogenesis. For example, Rhodes et al[109] performed meta-analysis of four independent prostate cancer microarray datasets and identified two important metabolic pathways in prostate cancer. As in silico modeling evolves to contain more complete cell signaling and enzymatic pathways information, it is imaginable if similar studies are conducted for other "omics" platforms, meta-signatures and regulation networks could potentially be deciphered.

In general, the development of robust bioinformatics analysis methods is very important and there are considerable rooms for improvement in the analysis of "omics" experiments. For example, for proteomics, due to the fact that the majority of published studies used SELDI approach, the data analysis mostly used software from Ciphergen. However, the Ciphergen software has been reported to be too conservative in peak finding and its baseline correction algorithm produces biased estimates of peak heights.[110] To remedy this, bioinformatics tools developed for microarray analysis should, in principle, complement proteomics analysis,[111] but are limited due to the unique attributes of proteomics datasets. A case in point is that the number of interrogations in microarrays is determined pre-experimentally by the number of genes/probes on the microarray chip. Thus the confidence in declaring some genes as differentially expressed can be quantified using statistical p values by calculating the rate of false identifications.[112] In contrast, proteomics analysis of complex mixtures like serum has no such a priori enumeration of targets and the discovery procedure is iterative and far more informal. It is therefore difficult to estimate the degree of confidence about any finding due to the lack of a described procedural structure at this time.[34]

Bioinformatics Applications in Nutritional Cancer Research

Advancement of bioinformatics tools and novel "omics" technologies offer enormous opportunity to investigate the highly complex relationship between nutrition and cancer.[113] In a recent report by the World Cancer Research Fund, ~40% of all cancers are directly linked to dietary nutrition.[114] Since there is a direct link between cancer incidence and nutrition, many studies are currently underway to identify and isolate important chemopreventive agents from natural nutritional souces.[115,116] An example is the γ-tocotrienol purified from palm oil that possesses selective anticancer property towards carcinogenic cells and no cytotoxic effects on normal cells.[115] The combination of various bioinformatics and "omics" technologies will greatly facilitate the discovery of new biomarkers associated with this phytonutrient and other dietary metabolites that affect carcinogenesis. It can be expected that the future "omics"-based human nutritional research can provide personalized dietary recommendations for cancer prevention and improvements in patient care.

References

1. Mooney S. Bioinformatics approaches and resources for single nucleotide polymorphism functional analysis. Brief Bioinform 2005; 6(1):44-56.
2. Larranaga P, Calvo B, Santana R et al. Machine learning in bioinformatics. Brief Bioinform 2006; 7(1):86-112.
3. Perez-Iratxeta C, Andrade-Navarro MA, Wren JD. Evolving research trends in bioinformatics. Brief Bioinform 2007; 8(2):88-95.
4. McLendon R, Friedman A, Bigner D et al. Comprehensive genomic characterization defines human glioblastoma genes and core pathways. Nature 2008; 455(7216):1061-1068.
5. Jemal A, Tiwari RC, Murray T et al. Cancer statistics. CA Cancer J Clin 2004; 54(1):8-29.
6. Bertucci F, Goncalves A. Clinical proteomics and breast cancer: strategies for diagnostic and therapeutic biomarker discovery. Future Oncol 2008; 4(2):271-287.
7. Richart RM. Current concepts in obstetrics and gynecology: The patient with an abnormal Pap smear—screening techniques and managment. N Engl J Med 1980; 302(6):332-334.
8. Otto SJ, Fracheboud J, Looman CW et al. Initiation of population-based mammography screening in Dutch municipalities and effect on breast-cancer mortality: a systematic review. Lancet 2003; 361(9367):1411-1417.
9. Simon JB. Fecal occult blood testing: clinical value and limitations. Gastroenterologist 1998; 6(1):66-78.
10. Mukoubayashi C, Yanaoka K, Ohata H et al. Serum pepsinogen and gastric cancer screening. Intern Med 2007; 46(6):261-266.
11. Takechi H, Ito K, Yamamoto T et al. Prostate-specific antigen kinetics in screen-detected prostate cancer in Japan. Urology 2008; 72(5):1111-1115.
12. Sturgeon C. Practice guidelines for tumor marker use in the clinic. Clin Chem 2002; 48(8):1151-1159.
13. Altman DG, Bland JM. Diagnostic tests. 1: Sensitivity and specificity. Bmj 1994; 308(6943):1552.
14. Fahey MT, Irwig L, Macaskill P. Meta-analysis of Pap test accuracy. Am J Epidemiol 1995; 141(7):680-689.
15. Pignone M, Rich M, Teutsch SM et al. Screening for colorectal cancer in adults at average risk: a summary of the evidence for the U.S. Preventive Services Task Force. Ann Intern Med 2002; 137(2):132-141.
16. Gloeckler Ries LA, Reichman ME, Lewis DR et al. Cancer survival and incidence from the Surveillance, Epidemiology and End Results (SEER) program. Oncologist 2003; 8(6):541-552.
17. Moore LE, Pfeiffer RM, Poscablo C et al. Genomic DNA hypomethylation as a biomarker for bladder cancer susceptibility in the Spanish Bladder Cancer Study: a case-control study. Lancet Oncol 2008; 9(4):359-366.
18. Shou J, Dotson C, Qian HR et al. Optimized blood cell profiling method for genomic biomarker discovery using high-density microarray. Biomarkers 2005; 10(4):310-320.

19. Purohit S, Podolsky R, Schatz D et al. Assessing the utility of SELDI-TOF and model averaging for serum proteomic biomarker discovery. Proteomics 2006; 6(24):6405-6415.
20. Dieterle F, Schlotterbeck G, Binder M et al. Application of metabonomics in a comparative profiling study reveals N-acetylfelinine excretion as a biomarker for inhibition of the farnesyl pathway by bisphosphonates. Chem Res Toxicol 2007; 20(9):1291-1299.
21. Cassidy F, Yatham LN, Berk M et al. Pure and mixed manic subtypes: a review of diagnostic classification and validation. Bipolar Disord 2008; 10(1 Pt 2):131-143.
22. Ames BN. Identifying environmental chemicals causing mutations and cancer. Science 1979; 204(4393):587-593.
23. Paez JG, Janne PA, Lee JC et al. EGFR mutations in lung cancer: correlation with clinical response to gefitinib therapy. Science 2004; 304(5676):1497-1500.
24. Venkitaraman AR. Cancer susceptibility and the functions of BRCA1 and BRCA2. Cell 2002; 108(2):171-182.
25. Jones PA, Baylin SB. The epigenomics of cancer. Cell 2007; 128(4):683-692.
26. Koch N, Tamalet C, Tivoli N et al. Comparison of two commercial assays for the detection of insertion mutations of HIV-1 reverse transcriptase. J Clin Virol 2001; 21(2):153-162.
27. Shendure J, Porreca GJ, Reppas NB et al. Accurate multiplex polony sequencing of an evolved bacterial genome. Science 2005; 309(5741):1728-1732.
28. Velculescu VE, Zhang L, Vogelstein B et al. Serial analysis of gene expression. Science 1995; 270(5235):484-487.
29. Jones S, Zhang X, Parsons DW et al. Core signaling pathways in human pancreatic cancers revealed by global genomic analyses. Science 2008; 321(5897):1801-1806.
30. Quackenbush J. Microarray analysis and tumor classification. N Engl J Med 2006; 354(23):2463-2472.
31. Fliser D, Wittke S, Mischak H. Capillary electrophoresis coupled to mass spectrometry for clinical diagnostic purposes. Electrophoresis 2005; 26(14):2708-2716.
32. Seibert V, Wiesner A, Buschmann T et al. Surface-enhanced laser desorption ionization time-of-flight mass spectrometry (SELDI TOF-MS) and ProteinChip technology in proteomics research. Pathol Res Pract 2004; 200(2):83-94.
33. Villas-Boas SG, Moxley JF, Akesson M et al. High-throughput metabolic state analysis: the missing link in integrated functional genomics of yeasts. Biochem J 2005; 388(Pt 2):669-677.
34. Boguski MS, McIntosh MW. Biomedical informatics for proteomics. Nature 2003; 422(6928):233-237.
35. Ginsburg GS, Haga SB. Translating genomic biomarkers into clinically useful diagnostics. Expert Rev Mol Diagn 2006; 6(2):179-191.
36. Goodacre R, Vaidyanathan S, Dunn WB et al. Metabolomics by numbers: acquiring and understanding global metabolite data. Trends Biotechnol 2004; 22(5):245-252.
37. Nicholson JK, Wilson ID. Opinion: understanding 'global' systems biology: metabonomics and the continuum of metabolism. Nat Rev Drug Discov 2003; 2(8):668-676.
38. Weckwerth W, Loureiro ME, Wenzel K et al. Differential metabolic networks unravel the effects of silent plant phenotypes. Proc Natl Acad Sci USA 2004; 101(20):7809-7814.
39. Soga T, Ohashi Y, Ueno Y et al. Quantitative metabolome analysis using capillary electrophoresis mass spectrometry. J Proteome Res 2003; 2(5):488-494.
40. Narod SA, Neuhausen S, Vichodez G et al. Rapid progression of prostate cancer in men with a BRCA2 mutation. Br J Cancer 2008; 99(2):371-374.
41. King MC, Marks JH, Mandell JB. Breast and ovarian cancer risks due to inherited mutations in BRCA1 and BRCA2. Science 2003; 302(5645):643-646.
42. Westra JL, Schaapveld M, Hollema H et al. Determination of TP53 mutation is more relevant than microsatellite instability status for the prediction of disease-free survival in adjuvant-treated stage III colon cancer patients. J Clin Oncol 2005; 23(24):5635-5643.
43. Okabe T, Okamoto I, Tsukioka S et al. Synergistic antitumor effect of S-1 and the epidermal growth factor receptor inhibitor gefitinib in nonsmall cell lung cancer cell lines: role of gefitinib-induced down-regulation of thymidylate synthase. Mol Cancer Ther 2008; 7(3):599-606.
44. Slater E, Amrillaeva V, Fendrich V et al. Palladin mutation causes familial pancreatic cancer: absence in European families. PLoS Med 2007; 4(4):e164.
45. Soegaard M, Kjaer SK, Cox M et al. BRCA1 and BRCA2 mutation prevalence and clinical characteristics of a population-based series of ovarian cancer cases from Denmark. Clin Cancer Res 2008; 14(12):3761-3767.
46. Guo CY, Xu XF, Wu JY et al. PCR-SSCP-DNA sequencing method in detecting PTEN gene mutation and its significance in human gastric cancer. World J Gastroenterol 2008; 14(24):3804-3811.
47. Setlur SR, Royce TE, Sboner A et al. Integrative microarray analysis of pathways dysregulated in metastatic prostate cancer. Cancer Res 2007; 67(21):10296-10303.

48. van de Vijver MJ, He YD, van't Veer LJ et al. A gene-expression signature as a predictor of survival in breast cancer. N Engl J Med 2002; 347(25):1999-2009.
49. van 't Veer LJ, Dai H, van de Vijver MJ et al. Gene expression profiling predicts clinical outcome of breast cancer. Nature 2002; 415(6871):530-536.
50. Resnick MB, Routhier J, Konkin T et al. Epidermal growth factor receptor, c-MET, beta-catenin and p53 expression as prognostic indicators in stage II colon cancer: a tissue microarray study. Clin Cancer Res 2004; 10(9):3069-3075.
51. Gordon GJ, Jensen RV, Hsiao LL et al. Translation of microarray data into clinically relevant cancer diagnostic tests using gene expression ratios in lung cancer and mesothelioma. Cancer Res 2002; 62(17):4963-4967.
52. Grutzmann R, Boriss H, Ammerpohl O et al. Meta-analysis of microarray data on pancreatic cancer defines a set of commonly dysregulated genes. Oncogene 2005; 24(32):5079-5088.
53. Mok SC, Chao J, Skates S et al. Prostasin, a potential serum marker for ovarian cancer: identification through microarray technology. J Natl Cancer Inst 2001; 93(19):1458-1464.
54. Yap YL, Zhang XW, Smith D et al. Molecular gene expression signature patterns for gastric cancer diagnosis. Comput Biol Chem 2007; 31(4):275-287.
55. Peter J, Unverzagt C, Krogh TN et al. Identification of precursor forms of free prostate-specific antigen in serum of prostate cancer patients by immunosorption and mass spectrometry. Cancer Res 2001; 61(3):957-962.
56. Ricolleau G, Charbonnel C, Lode L et al. Surface-enhanced laser desorption/ionization time of flight mass spectrometry protein profiling identifies ubiquitin and ferritin light chain as prognostic biomarkers in node-negative breast cancer tumors. Proteomics 2006; 6(6):1963-1975.
57. Shimma S, Sugiura Y, Hayasaka T et al. MALDI-based imaging mass spectrometry revealed abnormal distribution of phospholipids in colon cancer liver metastasis. J Chromatogr B Analyt Technol Biomed Life Sci 2007; 855(1):98-103.
58. Taguchi F, Solomon B, Gregorc V et al. Mass spectrometry to classify nonsmall-cell lung cancer patients for clinical outcome after treatment with epidermal growth factor receptor tyrosine kinase inhibitors: a multicohort cross-institutional study. J Natl Cancer Inst 2007; 99(11):838-846.
59. Ge G, Wong GW. Classification of premalignant pancreatic cancer mass-spectrometry data using decision tree ensembles. BMC Bioinformatics 2008; 9:275.
60. Yu J, Chen XW. Bayesian neural network approaches to ovarian cancer identification from high-resolution mass spectrometry data. Bioinformatics 2005; 21(Suppl 1):i487-494.
61. Ebert MP, Niemeyer D, Deininger SO et al. Identification and confirmation of increased fibrinopeptide a serum protein levels in gastric cancer sera by magnet bead assisted MALDI-TOF mass spectrometry. J Proteome Res 2006; 5(9):2152-2158.
62. Nithipatikom K, Borscheid CL, Kajdacsy-Balla A et al. Determination of cyclooxygenase and arachidonic acid metabolites in invasive human prostate cancer cells. Adv Exp Med Biol 2002; 507:269-274.
63. Ju YH, Fultz J, Allred KF et al. Effects of dietary daidzein and its metabolite, equol, at physiological concentrations on the growth of estrogen-dependent human breast cancer (MCF-7) tumors implanted in ovariectomized athymic mice. Carcinogenesis 2006; 27(4):856-863.
64. Marangon E, Sala F, Caffo O et al. Simultaneous determination of gemcitabine and its main metabolite, dFdU, in plasma of patients with advanced nonsmall-cell lung cancer by high-performance liquid chromatography-tandem mass spectrometry. J Mass Spectrom 2008; 43(2):216-223.
65. Ye B, Skates S, Mok SC et al. Proteomic-based discovery and characterization of glycosylated eosinophil-derived neurotoxin and COOH-terminal osteopontin fragments for ovarian cancer in urine. Clin Cancer Res 2006; 12(2):432-441.
66. Cheng D, Knox C, Young N et al. PolySearch: a web-based text mining system for extracting relationships between human diseases, genes, mutations, drugs and metabolites. Nucleic Acids Res 2008; 36(Web Server issue):W399-405.
67. Lawrence CE, Altschul SF, Boguski MS et al. Detecting subtle sequence signals: a Gibbs sampling strategy for multiple alignment. Science 1993; 262(5131):208-214.
68. Wood LD, Parsons DW, Jones S et al. The genomic landscapes of human breast and colorectal cancers. Science 2007; 318(5853):1108-1113.
69. Liu Y, Eyal E, Bahar I. Analysis of correlated mutations in HIV-1 protease using spectral clustering. Bioinformatics 2008; 24(10):1243-1250.
70. Bauer M, Schuster SM, Sayood K. The average mutual information profile as a genomic signature. BMC Bioinformatics 2008; 9:48.
71. Collyda C, Diplaris S, Mitkas P et al. Enhancing the quality of phylogenetic analysis using fuzzy hidden Markov model alignments. Stud Health Technol Inform 2007; 129(Pt 2):1245-1249.
72. Webb BJ, Liu JS, Lawrence CE. BALSA: Bayesian algorithm for local sequence alignment. Nucleic Acids Res 2002; 30(5):1268-1277.

73. Larsabal E, Danchin A. Genomes are covered with ubiquitous 11 bp periodic patterns, the "class A flexible patterns". BMC Bioinformatics 2005; 6:206.
74. Liu JJ, Cutler G, Li W et al. Multiclass cancer classification and biomarker discovery using GA-based algorithms. Bioinformatics 2005; 21(11):2691-2697.
75. Li L, Darden TA, Weinberg CR et al. Gene assessment and sample classification for gene expression data using a genetic algorithm/k-nearest neighbor method. Comb Chem High Throughput Screen 2001; 4(8):727-739.
76. Xiong M, Li W, Zhao J et al. Feature (gene) selection in gene expression-based tumor classification. Mol Genet Metab 2001; 73(3):239-247.
77. Zhang H, Yu CY, Singer B et al. Recursive partitioning for tumor classification with gene expression microarray data. Proc Natl Acad Sci USA 2001; 98(12):6730-6735.
78. Alon U, Barkai N, Notterman DA et al. Broad patterns of gene expression revealed by clustering analysis of tumor and normal colon tissues probed by oligonucleotide arrays. Proc Natl Acad Sci USA 1999; 96(12):6745-6750.
79. Crescenzi M, Giuliani A. The main biological determinants of tumor line taxonomy elucidated by a principal component analysis of microarray data. FEBS Lett 2001; 507(1):114-118.
80. Nguyen DV, Rocke DM. Tumor classification by partial least squares using microarray gene expression data. Bioinformatics 2002; 18(1):39-50.
81. Yap Y, Zhang X, Ling MT et al. Classification between normal and tumor tissues based on the pair-wise gene expression ratio. BMC Cancer 2004; 4:72.
82. Zhang XW, Yap YL, Wei D et al. Molecular diagnosis of human cancer type by gene expression profiles and independent component analysis. Eur J Hum Genet 2005; 13(12):1303-1311.
83. Robinson PN, Wollstein A, Bohme U et al. Ontologizing gene-expression microarray data: characterizing clusters with Gene Ontology. Bioinformatics 2004; 20(6):979-981.
84. Kim SY, Lee JW, Sohn IS. Comparison of various statistical methods for identifying differential gene expression in replicated microarray data. Stat Methods Med Res 2006; 15(1):3-20.
85. Petricoin EF, Belluco C, Araujo RP et al. The blood peptidome: a higher dimension of information content for cancer biomarker discovery. Nat Rev Cancer 2006; 6(12):961-967.
86. Qu Y, Adam BL, Yasui Y et al. Boosted decision tree analysis of surface-enhanced laser desorption/ ionization mass spectral serum profiles discriminates prostate cancer from noncancer patients. Clin Chem 2002; 48(10):1835-1843.
87. Hilario M, Kalousis A, Muller M et al. Machine learning approaches to lung cancer prediction from mass spectra. Proteomics 2003; 3(9):1716-1719.
88. Wagner S, Scholz K, Donegan M et al. Metabonomics and biomarker discovery: LC-MS metabolic profiling and constant neutral loss scanning combined with multivariate data analysis for mercapturic acid analysis. Anal Chem 2006; 78(4):1296-1305.
89. Wu B, Abbott T, Fishman D et al. Comparison of statistical methods for classification of ovarian cancer using mass spectrometry data. Bioinformatics 2003; 19(13):1636-1643.
90. Conrads TP, Fusaro VA, Ross S et al. High-resolution serum proteomic features for ovarian cancer detection. Endocr Relat Cancer 2004; 11(2):163-178.
91. Shin H, Markey MK. A machine learning perspective on the development of clinical decision support systems utilizing mass spectra of blood samples. J Biomed Inform 2006; 39(2):227-248.
92. Yang J, Xu G, Zheng Y et al. Diagnosis of liver cancer using HPLC-based metabonomics avoiding false-positive result from hepatitis and hepatocirrhosis diseases. J Chromatogr B Analyt Technol Biomed Life Sci 2004; 813(1-2):59-65.
93. Ntzani EE, Ioannidis JP. Predictive ability of DNA microarrays for cancer outcomes and correlates: an empirical assessment. Lancet 2003; 362(9394):1439-1444.
94. Simon R, Radmacher MD, Dobbin K et al. Pitfalls in the use of DNA microarray data for diagnostic and prognostic classification. J Natl Cancer Inst 2003; 95(1):14-18.
95. Efron B, Tibshirani R. Improvements on cross-validation: The .632+ bootstrap method. Journal of the American Statistical Association 1997; 92:548-560.
96. Adam BL, Qu Y, Davis JW et al. Serum protein fingerprinting coupled with a pattern-matching algorithm distinguishes prostate cancer from benign prostate hyperplasia and healthy men. Cancer Res 2002; 62(13):3609-3614.
97. Ein-Dor L, Zuk O, Domany E. Thousands of samples are needed to generate a robust gene list for predicting outcome in cancer. Proc Natl Acad Sci USA 2006; 103(15):5923-5928.
98. Ioannidis JP. Microarrays and molecular research: noise discovery? Lancet 2005; 365(9458):454-455.
99. Dodd L, Wagner R, Armato S et al. Lung Image Database Consortium Research Group. Assessment methodologies and statistical issues for computer-aided diagnosis of lung nodules in computed tomography: contemporary research topics relevant to the lung image database consortium. Acad Radiol 2004; 11(4):462-475.

100. Xu L, Tan AC, Naiman DQ et al. Robust prostate cancer marker genes emerge from direct integration of inter-study microarray data. Bioinformatics 2005; 21(20):3905-3911.

101. Jung S, Bang H, Young S. Sample size calculation for multiple testing in microarray data analysis. Biostatistics 2005; 6:157-169.

102. Pounds S, Cheng C. Sample size determination for the false discovery rate. Bioinformatics 2005; 21(23):4263-4271.

103. Wang SJ, Chen JJ. Sample size for identifying differentially expressed genes in microarray experiments. J Comput Biol 2004; 11(4):714-726.

104. Omenn GS, States DJ, Adamski M et al. Overview of the HUPO Plasma Proteome Project: results from the pilot phase with 35 collaborating laboratories and multiple analytical groups, generating a core dataset of 3020 proteins and a publicly-available database. Proteomics 2005; 5(13):3226-3245.

105. Rhodes DR, Yu J, Shanker K et al. ONCOMINE: a cancer microarray database and integrated data-mining platform. Neoplasia 2004; 6(1):1-6.

106. Rocca-Serra P, Brazma A, Parkinson H et al. ArrayExpress: a public database of gene expression data at EBI. C R Biol 2003; 326(10-11):1075-1078.

107. Sherlock G, Hernandez-Boussard T, Kasarskis A et al. The Stanford Microarray Database. Nucleic Acids Res 2001; 29(1):152-155.

108. Barrett T, Edgar R. Mining microarray data at NCBI's Gene Expression Omnibus (GEO)*. Methods Mol Biol 2006; 338:175-190.

109. Rhodes DR, Yu J, Shanker K et al. Large-scale meta-analysis of cancer microarray data identifies common transcriptional profiles of neoplastic transformation and progression. Proc Natl Acad Sci USA 2004; 101(25):9309-9314.

110. Coombes KR, Morris JS, Hu J et al. Serum proteomics profiling—a young technology begins to mature. Nat Biotechnol 2005; 23(3):291-292.

111. Kremer A, Schneider R, Terstappen GC. A bioinformatics perspective on proteomics: data storage, analysis and integration. Biosci Rep 2005; 25(1-2):95-106.

112. Benjamini Y, Hochberg Y. Controlling the false discovery rate: a practical and powerful approach to multiple testing. J R Statist Soc B 1995; 57:289-300.

113. Zhang X, Yap Y, Wei D et al. Novel omics technologies in nutrition research. Biotechnol Adv 2008; 26(2):169-176.

114. World Cancer Research Fund (WCRF) International and American Institute for Cancer Research (AICR). Food, nutrition and the prevention of cancer: a global perspective report commissioned by the World Cancer Research Fund and American Institute for Cancer Research Washington, DC: WCRF-AICR; 2003.

115. Sen CK, Khanna S, Roy S. Tocotrienols: Vitamin E beyond tocopherols. Life Sci 2006; 78(18):2088-2098.

116. Rimando AM, Suh N. Natural products and dietary prevention of cancer. Mol Nutr Food Res 2008; 52(Suppl 1):S5.

CHAPTER 11

Integrative Omics Technologies in Cancer Biomarker Discovery

Xuewu Zhang,* Lei Shi, Gu Chen and Yee Leng Yap

Abstract

In today's cancer management, there is a strong clinical imperative to identify disease-specific biomarkers to improve diagnosis, prognosis and to predict as well as monitor treatment efficiency. Advances in high-throughput omics technologies have vastly increased the space of prospective molecular biomarkers. Although each omics technology plays important roles in cancer research, different omics platforms have different strengths and limitations. The systematic integration of various omic data can prove advantageous for accelerating cancer biomarker discovery. In this chapter, we review the current applications of integrative omics technologies to cancer biomarker discovery.

Introduction

During the last two decades, cancer incidence is still increasing worldwide and the general ratio of deaths to new cancer cases remains as high as 49% overall, although the remarkable scientific and technological advances in medicine have been achieved.[1] Therefore, the discovery of more effective cancer therapeutics and diagnostics is urgently needed. Biomarkers are the key element of modern diagnostics and their value in medicine is ever increasing. To date, thanks to the availability of various omics-based technologies following the completion of Human Genome Project, the spectrum of current biomarker application is being rapidly expanded such that biomarkers can be used to diagnose disease risk or presence of disease in an individual, or to tailor treatments for the disease in an individual.

Recently, the technologies that measure some characteristic of a large family of cellular molecules, such as genes, proteins, or small metabolites, have been appended by the suffix "-omics", such as Genomics, Transcriptomics, Proteomics, Peptidomics, Metabonomics, Glycomics, Lipidomics, Phosphoproteomics, etc. The separate use of these omics technologies have extensively been investigated in the development of cancer biomarkers. In this chapter, we attempt to review the current application of integrative omics technologies to cancer biomarker discovery.

Currently Available Omics Technologies in Cancer Biomarker Discovery

Genomics

The term "genomics" refers to the study of all nucleotide sequences in the genome of an organism. The genomics in cancer biomarker discovery is to seek specific biomarkers related to genome alterations caused by cancer, for example, DNA sequence changes, copy number aberrations,

*Corresponding Author: XW Zhang—College of Light Industry and Food Sciences, South China University of Technology, 381 Wushan Road, Tianhe Area, Guangzhou 510640, China. Email: snow_dance@sina.com

Omics Technologies in Cancer Biomarker Discovery, edited by Xuewu Zhang. ©2011 Landes Bioscience.

chromosomal rearrangements and epigenetic modifications such as DNA methylation. The widely used genomic technologies in cancer research include single nucleotide polymorphism (SNP) array,[2] next-generation sequencing (NGS) technologies, such as Roche 454, ABI SOLiD, Illumina Solexa and Helicos.[3] Recently, DNA methylations have emerged as highly promising biomarkers and are being actively studied in multiple cancers. There are many advantages for DNA methylation biomarkers, for example, stability and easy detection using PCR or array-based approaches in blood, sputum, urine and stool, making it well suited for clinical noninvasive detection.[4]

Transcriptomics

Transcriptomics is to measure the relative amounts of all messenger RNAs (mRNAs) in a given organism for determining the patterns and levels of gene expression. A powerful technique used in transcriptomics is DNA microarray. Several commercial DNA microarray platforms are widely used, including Affymetrix, Illumina, Agilent, GE Healthcare and NCI_Operon. Recent uses of DNA microarrays for transcriptomic analysis in cancer research are not limited to the mRNA level and are also being extended to detect microRNAs (micro ribonucleic acids),[5] leading to the generation of a new member of omics, namely microRNAomics. Microarray-based gene expression profiling of human cancers has generated hundreds of novel diagnostic and prognostic biomarkers as well as therapeutic targets.[6,7]

Proteomics

Proteomics is the study of all proteins expressed in a cell, tissue, or organism, including all protein isoforms and posttranslational modifications. The proteomics approaches can be classified into two categories: gel-based and gel-free proteomics. For the gel-based proteomics experiments, proteins are separated and quantified by two-dimensional polyacrylamide gel electrophoresis (2D-Gel), with mass spectrometry to identify molecules of interest. Gel-free proteomics approach, shotgun proteomics, is involved in the combined use of multidimensional liquid chromatography (MDLC) combined with tandem mass spectrometry,[8] the basic strategy include: digesting proteins into peptides and sequencing them using tandem mass spectrometry and identifying them by automated database searching. Proteomics technology has been widely applied to analyses of tissue, serum, saliva, sputum, cerebrospinal fluid and nipple aspirate fluid in an attempt to identify biomarkers of cancer.[9]

Metabonomics

Metabonomics is to investigates the fingerprint of biochemical perturbations caused by disease, drugs and toxins. It is noted that another similar term "metabolomics" refers to comprehensive analysis of all metabolites generated in a given biological system, focusing on the measurements of metabolite concentrations and secretions in cells and tissues. Usually, metabonomics studies use urine or blood as samples. A number of methods can be used to produce metabolic signatures of biomaterials, including nuclear magnetic resonance (NMR), gas chromatography-mass spectrometry (GC-MS), liquid chromatography-mass spectrometry (LC-MS), capillary electrophoresis-mass spectrometry (CE-MS), Fourier transform ion cyclotron resonance mass spectroscopy (FTICR-MS) and ultra-performance liquid chromatography mass spectrometry system (UPLC-MS). Currently, there are more and more examples of metabonomics technologies being successfully used to cancer biomarker discovery.[10-13]

Peptidomics

Peptidomics is the simultaneous visualization and identification of the whole peptidome of a cell or tissue, that is, all expressed peptides with their posttranslational modifications. Usually, the candidate peptidomic biomarkers come either from the peptides and fragments derived from parental protein molecules or from the cleavage products generated, ex vivo, after blood clotting. Villanueva et al[14] developed an automated procedure for serum peptide profiling that utilizes magnetic, reverse-phase beads for peptides capture and MALDI-TOF-MS analysis. The results demonstrated that a pattern of 274 peptides can be used to correctly predict 96.4% of the samples with or without brain tumor. Recently, using a highly optimized peptide extraction and

MALDI-TOF-MS approach, Villanueva et al[15] profiled 106 serum samples from patients with advanced prostate cancer, bladder cancer and breast cancer and identified 61 signature peptides, which provides accurate class prediction between cancerous patients and controls. Surprisingly, the peptides identified as cancer-type-specific markers proved to be products of enzymatic breakdown generated after patient blood collection.

Glycomics

Glycosylation is a very important posttranslational modification of many biologically relevant molecules. Aberrant glycosylation occurs in essentially all types of experimental and human cancers. Glycomics is to identify and study all the glycan molecules produced by an organism, encompassing all glycoconjugates (glycolipids, glycoproteins, lipopolysaccharides, peptidoglycans and proteoglycans). On the other hand, a similar term is glycoproteomics, which refers only to the characterization of the glycosylation of proteins. Drake et al[16] has reviewed MS combined with lectin-based glycoprotein capture strategies for the discovery of serum glycoprotein biomarkers. Comparative studies of the specific carbohydrate chains of glycoproteins can provide useful information for the diagnosis, prognosis and immunotherapy of tumors.[17-20]

Phosphoproteomics

Phosphorylation of proteins is the most studied modification, which is directly involved in the regulation of many biological processes such as metabolism, transcription, translation, cell cycle progression, cytoskeletal rearrangement, cell movement, apoptosis and differentiation.[21] Phosphoproteomics is the characterization of the phosphorylation of proteins. Chalmers et al[22] provided an overview of the different methods/technologies currently available to identify protein phosphorylation sites. Reinders and Sickmann[23] reviewed the most frequently used methods in isolation and detection of phosphoproteins and -peptides. Preliminary studies showed that it is possible to mine potential cancer biomarkers from the tumor phosphoproteome.[24,25]

Lipidomics

Lipids are molecules that are highly soluble in organic solvents, the vast majority of naturally occurring lipids are small molecules (molecular mass <2,000 Da). Lipidomics is systems-level analysis and characterization of lipids and their interacting partners, can be viewed as a subdiscipline of metabonomics.[26] Umezu-Goto et al[27] reported that Lysophosphatidic acid (LPA), the simplest phospholipids, is markedly elevated in the ascetic fluid of ovarian cancer patients. Similarly, Wenk[26] found that the levels of sphingolipids are altered in various types of cancers. Recently, to facilitate the understanding of the role of lipid mediators in cancer, a LC/ESI-MS/MS assay was developed to perform lipidomic analysis of 27 mediators including prostaglandins, prostacyclines, thomboxanes, dihydroprostaglandins and isoprostanes.[28] Thus it is possible that a global analysis of lipid patterns could prove diagnostic for particular cancers.

Bioinformatics

Globally, the main similarity for all the omics technologies is that they all rely on analytical chemistry methods and generate complex datasets. Bioinformatics is to develop and apply data mining as well as machine learning algorithms involving mathematics, informatics, statistics, computer science, artificial intelligence, chemistry and physics to extract diagnostic or prognostic information from the complex data. Currently, bioinformatics has been widely applied to various omics technologies for cancer biomarker discovery. For example, the performance of different algorithms for transcriptomics investigations has been compared by Dudoit et al;[29] for proteomics experiments, the computational issues in the processing and classification of protein mass spectra have been reviewed in detail by Hilario et al[30] and the performance of various methods has been evaluated by Shin and Markey et al;[31] for metabonomic data, a few studies are available, such as a principle component analysis (PCA) method that was used for the diagnosis of liver cancer[32] and the statistical total correlation spectroscopy analysis method for biomarker identification from metabolic NMR datasets.[33]

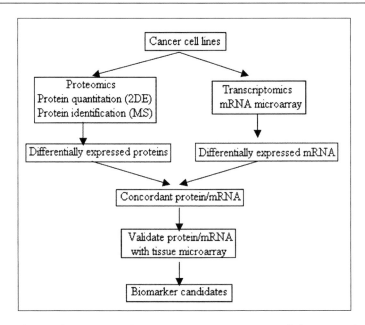

Figure 1. Scheme of an integrated proteo-transcriptomic approach for cancer biomarker discovery.

Integrative Use of Omics Technologies for Cancer Biomarker Discovery

To date, few reports about integrative use of omics technologies in the field of cancer biomarker discovery are available in literature, among limited applications are genomics, transcriptomics, proteomics and metabonomics.

Combination of Transcriptomics and Proteomics for Cancer Biomarker Discovery

To identify molecular markers for that differential diagnosis, Nishizuka et al[34] developed a multistep protocol starting with the 60 human cancer cell lines: (a) identification of candidate markers using cDNA microarrays; (b) verification of clone identities by resequencing; (c) corroboration of transcript levels using Affymetrix oligonucleotide chips; (d) quantitation of protein expression by "reverse-phase" protein microarray; and (e) prospective validation of candidate markers on clinical tumor sections in tissue microarrays. The two best candidate biomarkers were villin for colon cancer cells and moesin for ovarian cancer cells, which can be used as biomarkers to distinguish colon cancer from ovarian cancer. This is very important because the two cancers are all present in the abdomen and it is difficult to distinguish them, while correct diagnosis is essential because of their significantly different treatments. This study shows that the coupling of transcriptomics to proteomics has the potential to determine the origin of cancer tissue.

Ou et al[35] used a proteo-transcriptomic integrative strategy for discovering novel breast cancer biomarkers (Fig. 1). Initially, the comprehensive two-dimensional electrophoresis (2DE)/mass spectrometry (MS) proteomic maps of cancer (MCF-7 and HCC-38) and control (CCD-1059Sk) cell lines were generated and the differentially expressed cell-line proteins were then mapped to mRNA transcript databases of cancer cell lines and primary breast tumors to identify 9 candidate biomarkers that were concordantly expressed at the gene expression level.

Then ANX1 was reconformed and three other novel candidates, CRAB, 6PGL and CAZ2, were validated as differentially expressed proteins by immunohistochemistry on breast tissue microarrays. This study illustrates that the systematic integration of transcriptomics, proteomics and tissue microarray can prove advantageous for accelerating cancer biomarker discovery.

Wang et al[36] used integrative transcriptomics and proteomics for identification of novel liver cancer biomarkers. By tandem mass spectrometry the secretomes of 12 individual paired samples of liver cancer and adjacent normal tissues were analyzed and 1528 proteins were identified, among them 87 proteins in the hepatocellular carcinoma (HCC) group and 86 proteins in the normal group were found to show fivefold overexpression. Subsequently a novel paradigm in combining biomarkers that include an up-regulated cancer biomarker and a down-regulated organ-enriched marker was presented, finally, chitinase-3-like protein 1(CHI3L1) and mannan-binding lectin serine peptidase 2 (MASP2) were identified as the top biomarker pair for HCC diagnosis. The further ELISA assays were employed to evaluate this biomarker pair in a separate cohort of 25 serum samples of liver cancer patients and 15 age-matched normal controls, the result showed that the combined marker pair (CHI3L1/MASP2 ratio) performed better than either marker alone.

Combination of Transcriptomics and Metabonomics for Cancer Biomarker Discovery

Ippolito et al[37] employed a combination of transcriptomics and metabolomics to identify features of neuroendocrine (NE) cancers associated with a poor outcome. Firstly, they used GeneChip to generate a signature of 446 genes; secondly, this signature was used for in silico metabolic reconstructions of NE cell metabolism; thirdly, these reconstructions in turn were used to direct GC-MS/MS and LC-MS/MS analysis of metabolites in NE tumors and cell lines. Finally, a list of mRNA transcripts and metabolites indicative of a poor prognosis for various human NE cancers was provided, such as, dopa decarboxylase (DDC) was identified as a general biomarker for NE tumors and amiloride-binding protein 1 (ABP1) as a biomarker of poor-prognosis NE tumors. It is noted that this is a typical example showing how the hypothesis-driven approach is successfully used in metabonomic analysis, that is, gene signatures to generate in silico prediction for metabolic pathways, then to direct MS-based metabonomic analysis for identification of metabolites. Such an active method should be of higher importance, which can significantly decrease the number of MS experiments and the running time, financial expenses and increase the probability to be able to find the "fish", compared to aimless "fishing expedition" method.[38]

Combination of Proteomics and Bioinformatics for Cancer Biomarker Discovery

Bernal et al[39] developed an integrative functional informatics, which is based on the convergence and integration of proteomics, bioinformatics and highthroughput screening techniques, to accelerate the discovery and validation of novel biomarkers. This approach enables high-throughput testing of potential biomarkers without compromising high-specificity and sensitivity; hence it is an ideal tool for the validation of novel biomarkers.[40]

Gortzak-Uzan et al[41] described an integrated proteomic and bioinformatic analyses to identify putative biomarkers for ovarian cancer (Fig. 2). In this study, an in-depth proteomic analysis of selected biochemical fractions of human ovarian cancer ascites was conducted by multidimensional protein identification technology (MudPIT) and gel-enhanced LC-MS/MS. Subsequently, integrated bioinformatics analysis was used to map the ascites proteome data to several recently published proteomic data sets of human plasma, urine, 59 available ovarian cancer microarrays, as well as protein-protein interactions from the Interologous Interaction Database I²D (http://ophid.utoronto.ca/i2d), finally, a panel of 80 putative biomarkers and a protein-protein interaction network for 18 candidate biomarkers were identified.

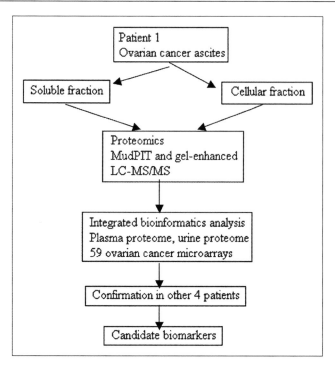

Figure 2. Scheme of an integrated proteomic and bioinformatics analyses for human ovarian cancer ascites to identify putative biomarkers.

Combination of Genomics, Transcriptomics and Proteomics for Cancer Biomarker Discovery

Li et al[42] applied an integrative omics approach (genomics, transcriptomics and proteomics) to the identification of the potential biomarkers in the diagnosis and therapy of lung cancer. Firstly, they identified 183 genes with increases in both genomic copy number and transcript in six lung cell lines. Secondly, 2D-gel electrophoresis and MS was used to identify 42 overexpressed proteins in the cancer cells relative to normal cells. Based on the comparison between the 183 genes and 42 proteins, four genes (PRDX1, EEF1A2, CALR and KCIP-1) were correlated with elevated protein expression. Following further validation experiments, the results showed that the amplification of EEF1A2 and KCIP-1 was associated with elevated protein expression, strongly suggesting that the two genes could be potential biomarkers for the diagnosis and therapy of lung cancer.

Conclusion

In the past few years, the principal focus of cancer biomarker research has been on the discovery of individual gene or protein biomarkers. However, genes/proteins do not operate as individual units; they collaborate in interconnected pathways to achieve different biological functions.[43] Once the perturbed pathways are known, it should be easier to monitor different aspects of cancer progression and therapy by focusing on pathways instead of individual genes/proteins. Recently, a new approach toward biomarker discovery, pathway-based biomarkers, has been proposed, in such a method pathways instead of individual proteins become the objective to be monitored and targeted.[44]

Dudley and butte[45] presented a novel framework for the identification of disease-specific protein biomarkers through the integration of biofluid proteomes and interdisease genomic relationships using a network paradigm. They created two biomarker networks: one is a blood plasma biomarker network by linking expression-based genomic profiles from 136 diseases to 1,028 detectable blood plasma proteins; the other is a urine biomarker network by linking genomic profiles from 127 diseases to 577 proteins detectable in urine. Through analysis of these molecular biomarker networks, they found that the majority (>80%) of putative protein biomarkers arc linked to multiple disease conditions. Thus, prospective disease-specific protein biomarkers are found in only a small subset of the biofluids proteomes, demonstrating the importance of considering shared molecular pathology across diseases including cancers when evaluating biomarker specificity.

Lawlor et al[46] presented a novel method for secretome analysis using "stacking gels", label-free relative quantitation and pathway analysis. With this technology, the secretome data obtained with six cancer cell lines from the three different tissues was uploaded into ingenuity pathways analysis (IPA). A total of 271 out of the 274-protein list predicted as either "secreted" or "plasma membrane" were identified and mapped to gene objects in the IPA Knowledge Base and used to generate biological networks. Then, functional analysis of the networks was done to identify the biological functions and/or diseases that were most significant to the genes in the network. The bioinformatics analysis suggested that several secretome proteins might be interconnected with intracellular canonical pathways, potentially opening the door for the use of secretomes to discover pathway-based biomarkers.

In summary, it is clear increasingly clear that, to fulfill the dream of personalized cancer care and treatment, we must move from the use of a single omic platform to the integration of multiple omic platforms and finally to systems biology, where the objective is to search pathway-based biomarkers rather than seek genomic/proteomic alterations of specific genes/proteins alone. However, a major challenge is that it is not easy to establish a direct link between genes and/or proteins and metabolites: multiple mRNAs could be formed from one gene; multiple proteins from one mRNA; multiple metabolites can be formed from one enzyme; and the same metabolite can participate in many different pathways. Usually, those genes that show minimal variation in their mRNA expression are more likely to show little or no correlation with the final protein levels, as the cell would likely control these proteins at the posttranslational level.[47] Similarly, a low correlation between protein and mRNA levels has also been described for a group of less abundant proteins.[48] Thus, the low correlation between the quantities of these molecules complicate the interpretation and integration of the various omics data. Anyway, it can be envisaged that the rapid development of new bioinformatics tools and integration of information from dynamic omics technologies (transcriptomics, proteomics and metabonomics, etc.) will certainly accelerate the discovery of novel pathway-based biomarkers.

References

1. Yeom YI, Kim SY, Lee HG et al. Cancer biomarkers in 'omics age. Biochip Journal 2008; 2(3):160-174.
2. Mao XY, Young BD, Lu YJ. The application of single nucleotide polymorphism microarrays in cancer research. Current Genomics 2007; 8(4):219-228.
3. von Bubnoff A. Next-generation sequencing: The race is on. Cell 2008; 132(5):721-723.
4. Anglim PP, Alonzo TA, Laird-Offringa IA. DNA methylation-based biomarkers for early detection of nonsmall cell lung cancer: an update. Molecular Cancer 2008; 7:81.
5. Schetter AJ, Leung SY, Sohn JJ et al. MicroRNA expression profiles associated with prognosis and therapeutic outcome in colon adenocarcinoma. JAMA 2008; 299:425-36.
6. Potti A, Dressman HK, Bild A et al. Genomic signatures to guide the use of chemotherapeutics. Nat Med 2006; 12:1294-300.
7. Schlabach MR, Luo J, Solimini NL et al. Cancer proliferation gene discovery through functional genomics. Science 2008; 319:620-624.
8. Washburn MP, Wolters D, Yates JR. Large-scale analysis of the yeast proteome by multidimensional protein identification technology. Nat Biotechnol 2001; 19:242-247.
9. Zhang XW, Wei D, Yap YL et al. Mass spectrometry-based "omics" technologies in cancer diagnostics. Mass Spectrometry Reviews 2007; 26:403-431.
10. Yang J, Xu G, Zheng Y et al. Diagnosis of liver cancer using HPLC-based metabonomics avoiding false-positive result from hepatitis and hepatocirrhosis diseases. J Chromatogr B Analyt Technol Biomed Life Sci 2004; 813(1-2):59-65.

11. Odunsi K, Wollman RM, Ambrosone CB et al. Detection of epithelial ovarian cancer using 1H-NMR-based metabonomics. Int J Cancer 2005; 113(5):782-788.

12. Al-Saffar NMS, Troy H, de Molina AR et al. Noninvasive magnetic resonance spectroscopic pharmaco-dynamic markers of the choline kinase inhibitor MN58b in human carcinoma models. Cancer Research 2006; 66(1):427-434.

13. Seidel A et al. Modified nucleosides: an accurate tumor marker for clinical diagnosis of cancer, early detection and therapy control. British Journal of Cancer 2006; 94(11):1726-1733.

14. Villanueva J, Philip J, Entenberg D et al. Serum peptide profiling by magnetic particle-assisted, automated sample processing and MALDI-TOF mass spectrometry. Anal Chem 2004; 76(6):1560-1570.

15. Villanueva J, Shaffer DR, Philip J et al. Differential exoprotease activities confer tumor-specific serum peptidome patterns. J Clin Invest 2006; 116(1):271-284.

16. Drake RR et al. Lectin capture strategies combined with mass spectrometry for the discovery of serum glycoprotein biomarkers. Mol Cell Proteomics 2006; 5:1957-1967.

17. Marrero JA, Lok AS. Newer markers for hepatocellular carcinoma. Gastroenterology 2004; 127 (5 Suppl 1):S113-S119.

18. Block TM, Comunale MA, Lowman M et al. Use of targeted glycoproteomics to identify serum glycoproteins that correlate with liver cancer in woodchucks and humans. Proc Natl Acad Sci USA 2005; 102(3):779-784.

19. Kobata A, Amano J. Altered glycosylation of proteins produced by malignant cells and application for the diagnosis and immunotherapy of tumours. Immunol Cell Biol 2005; 83:429-439.

20. Comunale MA et al. Proteomic analysis of serum-associated fucosylated glycoproteins in the development of primary hepatocellular carcinoma. J Proteome Res 2006; 5:308-315.

21. Raggiaschi R, Gotta S, Terstappen GC. Phosphoproteome analysis. Biosci Rep 2005; 25(1-2):33-44.

22. Chalmers MJ, Kolch W, Emmett MR et al. Identification and analysis of phosphopeptides. J Chromatogr B Anal Technol Biomed Life Sci 2004; 803(1):111-120.

23. Reinders J, Sickmann A. State-of-the-art in phosphoproteomics. Proteomics 2005; 5(16):4052-4061.

24. Kim JE, Tannenbaum SR, White FM. Global phosphoproteome of HT-29 human colon adenocarcinoma cells. J Proteome Res 2005; 4(4):1339-1346.

25. Lim YP. Mining the tumor phosphoproteome for cancer markers. Clin Cancer Res 2005; 11(9):3163-3169.

26. Wenk MR. The emerging field of lipidomics. Nat Rev Drug Discov 2005; 4(7):594-610.

27. Umezu-Goto M, Tanyi J, Lahad J et al. Lysophosphatidic acid production and action: Validated targets in cancer? J Cell Biochem 2004; 92(6):1115-1140.

28. Masoodi M, Nicolaou A. Lipidomic analysis of twenty-seven prostanoids and isoprostanes by liquid chromatography/electrospray tandem mass spectrometry. Rapid Commun Mass Spectrom 2006; 20(20):3023-3029.

29. Dudoit S et al. Comparison of discrimination methods for the classification of tumors using gene expression data. J Am Stat Assoc 2002; 97:77-87.

30. Hilario M et al. Processing and classification of protein mass spectra. Mass Spectrom Rev 2006; 25:409-449.

31. Shin H, Markey MK. A machine learning perspective on the development of clinical decision support systems utilizing mass spectra of blood samples. J Biomed Inform 2006; 39:227-248.

32. Yang J et al. Diagnosis of liver cancer using HPLC-based metabonomics avoiding false-positive results from hepatitis and hepatocirrhosis diseases. J Chromatogr B: Analyt Technol Biomed Life Sci 2004; 813:59-65.

33. Cloarec O et al. Statistical total correlation spectroscopy: an exploratory approach for latent biomarker identification from metabolic 1HNMR data sets. Anal Chem 2005; 77:1282-1289.

34. Nishizuka S, Chen ST, Gwadry FG et al. Diagnostic markers that distinguish colon and ovarian adenocarcinomas: Identification by genomic, proteomic and tissue array profiling. Cancer Res 2003; 63(17):5243-5250.

35. Ou K, Yu K, Kesuma D et al. Novel breast cancer biomarkers identified by integrative proteomic and gene expression mapping. Journal of Proteome Research 2008; 7(4):1518-1528.

36. Wang J, Gao F, Mo F et al. Identification of CHI3L1 and MASP2 as a biomarker pair for liver cancer through integrative secretome and transcriptome analysis. Proteomics Clinical Applications 2009; 3(5): 541-551.

37. Ippolito JE, Xu J, Jain S et al. An integrated functional genomics and metabolomics approach for defining poor prognosis in human neuroendocrine cancers. Proc Natl Acad Sci USA 2005; 102(28):9901-9906.

38. Kell DB. Metabolomics, modelling and machine learning in systems biology—towards an understanding of the languages of cells. FEBS J 2006; 873-894.

39. Bernal A et al. Emerging opportunities in functional informatics. Pharma Genomics 2003; 9:38-44.

40. Llyin SE et al. Biomarker discovery and validation: technologies and integrative approaches. Trends Biotechnol 2004; 22:411-416.
41. Gortzak-Uzan L, Ignatchenko A, Evangelou AI et al. A proteome resource of ovarian cancer ascites: Integrated proteomic and bioinformatic analyses to identify putative biomarkers. Journal of Proteome Research 2008; 7(1):339-351.
42. Li R et al. Identification of putative oncogenes in lung adenocarcinoma by a comprehensive functional genomic approach. Oncogene 2006; 25:2628-2635.
43. Vogelstein B, Kinzler KW. Cancer genes and the pathways they control. Nat Med 2004; 10(8):789-99.
44. Sawyers CL. The cancer biomarker problem. Nature 2008; 452(7187):548-552.
45. Dudley JT, Butte AJ. Identification of discriminating biomarkers for human disease using integrative network biology. Pacific Symposium on Biocomputing 2009;27-38.
46. Lawlor K, Nazarlan A, Lacomis L et al. Pathway-based biomarker search by high-throughput proteomics profiling of secretomes. Journal of Proteome Research 2009; 8(3):1489-1503.
47. Greenbaum D, Colangelo C, Williams K et al. Comparing protein abundance and mRNA expression levels on a genomic scale. Genome Biology 2003; 4:117.
48. Gygi SP, Rist B, Gerber SA et al. Quantitative analysis of complex protein mixtures using isotope-coded affinity tags. Nat Biotechnol 1999; 17:994-999.

Index

M

Mass spectrometry (MS) 31, 40-45, 50-53,
   55-57, 61-75, 83, 84, 86, 87, 89, 94, 96,
   101, 102, 107, 108, 119-122, 130-134
Metabolomics 50, 119, 130, 133
Metabonomics 1, 15, 16, 49-53, 118, 119,
   121, 129-133, 135
Methylation 1-7, 9-16, 41, 64-66, 85, 118,
   130
Microarray 4, 5, 7, 23-33, 42, 45, 62, 72, 95,
   97, 119, 120, 122, 130, 132, 133
Multiple Affinity Protein Profiling
   (MAPPing) 83, 85

N

Neoplasm 104, 105
Next generation sequencing (NGS) 1, 3, 6,
   15, 130
N-glycosylation 90
Nuclear magnetic resonance (NMR) 50-53,
   107, 108, 119, 130, 131

O

O-glycosylation 90

P

Peptidomics 55-58, 129, 130
Personalized healthcare 49
Phosphoproteomics 44, 61, 62, 68, 70, 72,
   74-76, 129, 131
Point mutation 13, 74
Prognosis 1, 2, 10-14, 23, 24, 33, 45, 46, 49,
   52, 53, 56, 63, 89, 104-106, 121, 129,
   131, 133
Prognostic biomarker 40, 45, 89, 120, 130
Protein identification 40, 42, 56, 62, 94,
   119, 133
Proteomics 1, 15, 16, 39-46, 49, 52, 53,
   55-58, 62, 65, 72, 74, 75, 83, 84, 89, 94,
   96, 109, 119-123, 129-135

S

Serological proteome analysis (SERPA) 83,
   84
Single nucleotide polymorphism (SNP) 1-3,
   6, 8, 9, 13-15, 118, 130
Survival 2, 12, 14, 32, 44-46, 49, 61, 63, 92,
   118, 120

T

Therapeutics 1, 9, 14, 16, 23, 24, 31, 39, 40,
   43-46, 49, 61-64, 75, 89, 117, 118, 122,
   129, 130
Therapeutic target 31, 44, 45, 122, 130
Therapy 2, 13, 15, 24, 31, 44-46, 49, 52, 55,
   61-64, 75, 89, 101, 105-107, 117, 118,
   120, 121, 134
Tocotrienol 123
Transciptome 24
Transcriptomics 1, 15, 16, 23-25, 33, 49, 52,
   118-121, 129-135
Tumor 1, 2, 4, 6, 8-13, 14, 16, 25, 32, 33,
   39, 43-46, 51, 52, 57, 58, 61-63, 75, 83,
   90, 92, 101, 104-106, 118, 119, 121,
   130-133

V

Validation 1, 30-33, 40, 45, 46, 53, 58, 61,
   67, 69, 70, 72, 75, 96, 97, 118, 121, 122,
   132-134